John Allen Paulos

BEYOND NUMERACY

John Allen Paulos received his Ph.D. in mathematics from the University of Wisconsin. Now professor of mathematics and presidential scholar at Temple University in Philadelphia, he is the author of the widely acclaimed books *Mathematics and Humor* and *I Think, Therefore I Laugh,* and, most recently, of the national best-seller *Innumeracy: Mathematical Illiteracy and Its Consequences.*

BEYOND NUMERACY

ALSO BY JOHN ALLEN PAULOS

Innumeracy: Mathematical Illiteracy and Its Consequences

I Think, Therefore I Laugh

Mathematics and Humor

Beyond NUMERACY

∞

RUMINATIONS OF A NUMBERS MAN

JOHN ALLEN PAULOS

VINTAGE BOOKS
A DIVISION OF RANDOM HOUSE, INC.
NEW YORK

FIRST VINTAGE BOOKS EDITION, APRIL 1992

Published in the United States by Vintage Books, a division of Random House, Inc., New York, and simultaneously in Canada by Random House of Canada Limited, Toronto. Originally published in hardcover by Alfred A. Knopf, New York, in 1991.

Library of Congress Cataloging-in-Publication Data
Paulos, John Allen.
Beyond numeracy: ruminations of a numbers man/John Allen Paulos—
1st Vintage Books ed.
p. cm.
Includes index.
ISBN 0-679-73807-X (pbk.)
1. Mathematics. I. Title.
QA7.P28 1992
510—dc20 91-50598
CIP

Manufactured in the United States of America
B987654321

For my parents, Helen and Peter,
the source of my X's and Y's

I'd also like to thank Rafe Sagalyn,
Robert Frankel, and Sheila, Leah,
and Daniel Paulos.

CONTENTS

INTRODUCTION

This book is in part a dictionary, in part a collection of short mathematical essays, and in part the ruminations of a numbers man. Although it contains many entries (brief essays) arranged in alphabetical order and depicting a broad range of mathematical topics, the book differs from a standard dictionary in that its entries are less comprehensive, longer, and in some cases quite unconventional.

By necessity, this book contains more facts and principles than most essay compilations. I have nevertheless tried to retain the personal tone and unifying voice of such collections. In other words, this book was written by an individual with specific interests (not all of them mathematical), biases (math as a liberal art and not merely as a tool for technologists), and pedagogical strategies (involving storytelling and unusual applications). Although the subject is mathematics and not me, I have made no attempt to exclude myself from the picture, hoping to serve as the reader's personal guide through a subject that is unnecessarily daunting to many. My intended audience is intelligent and literate, but largely innumerate (mathematically illiterate).

I received a surprising volume of mail from readers of my previous book *Innumeracy* who wrote that the book had whetted their interest in mathematics, and that they now wanted something more to satisfy their newly awakened appetite for the subject—something of similar style and texture that would take them beyond simple numeracy. I immodestly quote from one woman's letter: "It's innumerate to say perhaps, but I wish you would write another book that was exactly the same, only different—something that goes a little further." I hope this book will be both appealing and helpful to her and successful in providing its readers with a nontechnical but mostly uncompromising view of mathematics and its relation to our world.

There are a lot of people who appreciate the beauty and importance of mathematics but who, short of going back to college, see no way to pursue this interest. They've been led to believe that without a knowledge of formalisms, theorems, and symbolic manipulations, mathematical ideas are utterly beyond them. I think this is false and even pernicious. One can learn from Montaigne, Flaubert, and Camus without reading French, and one can learn from Euler, Gauss, and Gödel without solving differential equations. What's needed in each case is a translator fluent in both languages.

As such a would-be translator, I've striven to avoid as much as possible not only equations but also elaborate diagrams, tables, and formal symbols. I do include a few illustrations and brief mentions of some common mathematical notations because these are sometimes indispensable and especially useful when consulting other books. For the most part, however, the exposition uses words—English ones.

The entries range from summaries of whole disciplines (calculus, trigonometry, topology) to biographical and historical asides (Gödel, Pythagoras, non-Euclidean geometry) to bits of mathematical or quasi-mathematical folklore (infinite sets, Platonic solids, QED) well known to mathematicians but not to the educated layman and laywoman. Occasionally, I include less conventional pieces—a review of a non-existent book, a stream-of-mathematical-consciousness car trip, brief discussions of humor or ethics. New areas are discussed (chaos and fractals, recursion, complexity) as well as more classical ones (conic sections, mathematical induction, prime numbers).

I'm guilty throughout of committing flagrant "category mistakes": including as entries mathematical topics, pedagogical principles, little homilies, anecdotal vignettes as if they were all coordinate. I don't apologize, since these disparate discussions illustrate the frequently overlooked fact that mathematics is a many-layered human endeavor and not just a body of formal theorems and calculations.

Writing mathematical papers is different from writing about mathematics, but I think there needn't and shouldn't be such a chasm between the two activities. (I've often daydreamed of announcing the solution to a famous unsolved problem in a popular book instead of in a traditional research journal.) With regard to the accuracy of the various entries, I've tried here to steer a delicate course: writing precisely enough to preclude collegial disdain (collegial disinterest in such a popular work is inevitable), yet clearly enough to avoid readers' possible misconceptions. When clarity and precision come into conflict, as they sometimes do, I've opted in most cases for clarity.

A widespread misconception about mathematics is that it is completely hierarchical—first arithmetic, then algebra, then calculus, then more abstraction, then whatever. (What comes after advanced calculus? Answer: serious gum disease.) This belief in the totem pole nature of mathematics isn't true, but it prevents many people who did poorly in seventh-grade, high school, or even college mathematics from picking up a popular book on the subject. Often very "advanced" mathematical ideas are more intuitive and comprehensible than are certain areas of elementary algebra. My point is: If you get stuck or don't understand something, continue on and the fog will probably lift, often even in the same entry.

Finally, I recall the self-described innumerates I've known who have been surprised when I've remarked on their mathematical insight. Having a traditional computational view of mathematics, such people usually characterize their insightful comments as logic or common sense and not as mathematics; they remind me of the Molière character who was shocked to discover that he had been speaking prose all his life. This book is therefore written for unknowing mathophiles (among others) who have been thinking math all their lives without realizing it.

The entries are largely independent and lightly cross-referenced.

BEYOND NUMERACY

A MATHEMATICAL ACCENT

∞

The practical importance of studying mathematics is widely acknowledged, but relatively few will grant that the mathematics of everyday life can be an engaging topic for idle thought. Mathematics, however, provides a way of viewing the world, and developing a mathematical consciousness or outlook can enhance our daily rounds.

Rather than argue for this, let me illustrate with an extended anecdote. Recently I had to drive to New York on short notice and I was a little late. Behind a line of cars on the way to the turnpike, I developed the usual murderous thoughts as I noted that the driver in the lead car was letting cars from a busy side street turn right onto the street before him (and me). There was a traffic light at the intersection, so there was no need for this bit of philanthropy, and the would-be samaritan should have balanced his good deed against the aggravation to the drivers behind him. The mathematical integral or sum of the latter inconveniences was in this case greater. Although not a deep thought by any means, it and similar "calculations" seem to be quite foreign to many.

Finally reaching the expressway, I quickly accelerated to the prevailing speed of approximately 70 miles per hour, slowing down to something near 55 only when patrol cars appeared. Despite my rush, the inanity of the game seemed especially stark that day, and I wondered why no one had ever implemented the following simpleminded idea for curbing excessive speed on toll roads: When someone enters such a road, the ticket he or she receives is stamped with the time of entrance. The distance between the various toll booths is known, so that when the exit time is recorded later by a computer, it can easily calculate the person's average velocity while on the toll road. The toll booth operator may then direct speeders with incriminating toll tickets to a waiting patrol car.

The method wouldn't do away with all speeding, of course, since someone could still speed until just before the exit, stop for a cup of coffee or a full meal if they'd really been racing, and exit with a legal average velocity. Still the primary inducement for speeding would be gone. What's wrong with this plan? Division of one number by another, the distance traveled by the time elapsed, is surely not a risky or a novel technology. Speeding tickets are presently issued on the basis of radar, which is considerably less reliable.

Turning on the radio to escape the topic, I was reminded how I would so like to hear just once a rock song that uses the word "doesn't" rather than "don't" as in "She don't love me anymore" or, what was actually playing, "It don't matter anyway." Possibly because of the relative sensory deprivation of driving, the latter lugubrious lyric stuck with me. Maybe it didn't matter anyway, and if so, I wondered if it mattered that it didn't matter. If nothing mattered and that nothing mattered didn't matter either, then why couldn't we iterate? It didn't matter that it didn't matter that nothing mattered. And so on recursively.

I inhaled the fumes of the New Jersey Turnpike and further considered the situation. If nothing mattered, but that nothing mattered did matter, then we would have a rather dispiriting situation. If nothing mattered and that nothing mattered didn't matter either, then we would have the possibility of something better—an ironic and conceivably happy approach to life. Similarly at higher levels. Reasoning formally and probably simplistically, the best situation would be for things to matter at the basic level or, failing that, to matter on no level—either complete childhood simplicity or total adult irony. (See the entry on *time*.)

As I neared the Hess oil refinery, my thoughts turned to writing and publishing, but my absurdist mood persisted. Was there, given the large and increasingly homogeneous reading population, less of a "need" today for writers? Assuming that people read about the same number of books, magazines, and newspapers as they always did, and that they wanted to read the "best" of whatever they did read (as determined by best-seller lists, say), and that they tended, in large part, to read material written by their countrymen, it seemed to follow that the larger the nation, the smaller the percentage of its citizens who could be authors or, what one day might become equivalent, best-selling authors.

I thought of several counter-arguments, the most interesting pointing to the larger variety of publications (especially nonfiction books, magazines, and newsletters) that catered to ever more specialized tastes and that provided more opportunities for writers. If there was any substance to these vague musings, the likelihood of attaining literary stardom was shrinking, while the chances of earning a living with one's word processor were rising.

The radio was reporting a one-hour delay in the Lincoln Tunnel, so I decided to take the George Washington Bridge into Manhattan. This turned out to be not much better, since the victims of a minor accident were on the side of the road attracting the usual rubberneck response from passing motorists. The cumulative effect of everyone's slowing down to see that there was really nothing to see struck me as a miniaturized version of many human problems. No evil intent, just a common impulse whose magnification was unpleasant.

Traffic cleared after a twenty-minute delay only to clot up again even worse due to construction. A single-lane stretch of about a mile before the bridge was needlessly peppered with Do Not Pass signs. The signs reminded me of progressive sentences in which each succeeding word is one letter longer than its predecessor, and I whiled away the time dilating on the pattern. Finally I came up with "I Do Not Pass Since Danger Expands Anywhere Unmindful Speedsters Proliferate Unmanageably," of which I was inordinately proud.

Tiring of this, I noted the increasing frequency of MD license plates as we approached New York and remembered the statistic I had just read that there are 428 physicians in all of Ethiopia—a country of 40,000,000 people whose average life expectancy was under 40. Trying to keep my impatience at bay, I constructed biographies of people from their vanity license plates and concluded without a shred of evidence that I was right every time. This brought to mind the punch line of a license-plate joke told me by a mathematician friend: How do you spell the name "Henry"? Answer: H-E-N-3-R-Y—the 3 is silent.

On the bridge I remembered that the supporting cables take the form of a curve called a catenary unless weights are hung from them at equal intervals, in which case the shape is parabolic. I considered the likelihood of the bridge's collapse and then the still unlikely, but vastly more probable possibility of being killed by a drunk driver, or ultimately contracting cancer from my repeated exposure to the New Jersey Turnpike, or developing high blood pressure from the frustration of being locked in a car alone with my obsessive ruminations.

I arrived in New York only five minutes late for my appointment, but that isn't the point of the story. Its purpose was to illustrate a mathematical stream of consciousness. The topics which arose naturally were social trade-offs (the good samaritan, gawking at accidents), average velocity (speeding tickets on the turnpike), the logical level of statements (the "nothing matters" business), probability (chances of becoming a published author), wordplay (snowball sentences), and estimation (dying from a bridge collapse vs. other, more likely demises).

For most nonscientists, what's most important in science education is not the imparting of any particular set of facts (although I don't mean to denigrate factual knowledge), but the development of a scientific habit of mind: How would I test that? What's the evidence for it? How does this relate to other facts and principles? The same, I think, holds true in mathematics education. Remembering this formula or that theorem is less important for most people than is the ability to look at a situation quantitatively, to note logical, probabilistic, and spatial relationships, and to muse mathematically. (See the entry on *computation and rote.*)

Of course, I'm not advocating an exclusive focus on such musings, just a realization that mathematics is much more than computation, that the outlook that results from studying it can illuminate aspects of our lives other than our financial or scientific concerns. At the very least it can provide an alternative way to fill our driving time.

ALGEBRA—

SOME BASIC PRINCIPLES

∞

I always liked elementary algebra partly because my first teacher of it had been recruited (dragooned might be a better word) into teaching the subject even though she didn't know a quadratic formula from an upper leg exercise regimen. Being honest, however, she didn't try to hide this fact, and being near retirement, she depended on her better students to help her out of any mathematical difficulties. She often found some pretext to have one of us come to her classroom after school, at which time she always managed to rehearse the next day's lesson. Happily (and necessarily) she stressed a few basic principles and left most of the details to us.

Despite knowing some mathematics, I'll try to follow her sound pedagogical example of sketching the broad picture and avoiding technicalities as much as possible. This is especially important in algebra, whose mere mention brings back for many people miserable memories of trying to determine Henry's age when told that he is 5 times as old as his son but will in 4 years' time be only 3 times as old. There are many reasons for this distaste, but I sometimes wonder if one of them might not be traceable to algebra's discovery by Al-Khowarizmi in the early ninth century. This Al-Khowarizmi, from whose name the English word "algorithm" is derived and from whose influential book *Al-jabr wa'l Muqabalah* we get our word "algebra," was one of the preeminent mathematicians of a most impressive era in Arabic learning. His book deals with the solving of various simple sorts of equations, but true to what algorithm has come to mean, Al-Khowarizmi concentrated almost exclusively on recipes, formulas, rules, and procedures. To my mind, his text has little of the elegance or logical appeal of Euclid's *Elements,* but like the latter it was the standard work in its field for a very long time.

Although Al-Khowarizmi did not use variables in his algebra problems for the very good reason that they would not be invented for another 750 years, elementary algebra has come to be considered a generalization of arithmetic in which variables are used to stand for unknown numbers. (See the entry on *variables.*) This provides for vastly greater scope since, for example, the distributive law of numbers may be expressed by the single equation $A(X + Y) = AX + AY$. In arithmetic only specific instances of this law may be cited: $6(7 + 2) = (6 \times 7) + (6 \times 2)$, or $11(8 + 5) = (11 \times 8) + (11 \times 5)$. (Let me interject here a common little puzzle whose solution depends on the distributive law. Pick a whole number X. Add 3 to it. Double what you get. Subtract 4 from this result. Subtract twice your original number. Add 3. The answer must always be 5. Why?)

The title *Al-jabr wa'l Muqabalah* means something like "restoration and balancing" and refers to the basic algebraic insight that the solving of equations requires that one "balance" both sides of the equation, that if one performs an operation on one side of the equation, then one must "restore" equality by performing this same operation on the other side as well.

A review of this process appears later, but since you probably don't care that Henry is 20 and became a father at 16, consider the following more practical problem which arose out of the labyrinthine complexity of my financial empire. I was trying to decide recently where to put some money for a short term of 3 months—into a fund that paid 9% for the first month and then reverted to whatever the treasury bill rate would be for the next two months, or into a fund that paid 5.3% tax-free. I wondered what the average treasury bill rate, R, would have to be for the second and third month in order to break even with the fund's tax-free rate. This led to the equation $.72[(.09 + R + R)/3] = .053$, the .72 reflecting my 28% tax bracket. Similar equations arise in many areas of business, science, and everyday life, and the simple techniques used in solving them enabled me to find $R = 6.54\%$, far enough below the then treasury bill rate that I put my money into the taxed fund.

Algebra deals also with methods for solving quadratic equations ($X^2 + 5X + 3 = 0$, for example), cubic equations ($X^3 + 8X^2 - 5X + 1 = 0$), and equations of higher degree ($X^N + 5X^{(N-1)} \ldots - 11.2X^3 + 7$) as well as with equations involving two or more variables and with collections of such equations. (See the entries on *the quadratic and other formulas* and *linear programming.*) In all these, variants and refinements of Al-Khowarizmi's fundamental principle of "restoring and balancing" are employed, as is the crucial understanding that the variables stand for numbers and hence manipulations of them must be governed by the same arithmetic rules that govern numbers.

There is another mathematical discipline that goes by the name of algebra (sometimes called abstract algebra to distinguish it from elementary algebra), but its logic, historical roots, and flavor are sufficiently different that I'll reserve discussion of it to the entry on *groups.* Algebraists whose specialty is the study of abstract algebraic structures are often slightly chagrined when nonmathematicians assume that they're busy solving quadratic equations.

[Derivation of Henry's age from the fact that he's 5 times as old as his son but 4 years hence will be only 3 times as old: Start by supposing X is the present age of Henry's son. (I have a friend who claims to have begun an explanation in this way only to have someone at the back of the class respond with a jeering "Yeah, but suppose X ain't his age.") Then 5X must be Henry's present age. In 4 years the son will be ($X + 4$) years old, while Henry will be ($5X + 4$) years old. We're told that at this point the father will be only 3 times as old as his son. Stated algebraically, the relation *translates* into the equation $5X + 4 = 3(X + 4)$. Here we must use the distributive law to write $3(X + 4)$ as $3X + 12$. Thus $5X + 4 = 3X + 12$. Now we do our restoring and balancing. To preserve equality and simplify the equation, we subtract 4 from both sides of it and obtain $5X = 3X + 8$. For the same reasons we subtract 3X from both sides of this equation and get $2X = 8$. Finally we divide both sides by 2 to find that $X = 4$. Since X is the son's age, we conclude further that $5X = 20$ is Henry's age.

[The solution to the other puzzle: If X is the original number, then the subsequent transformations of it are ($X + 3$); $2(X + 3)$ or ($2X + 6$); ($2X + 2$); 2; 5.]

ANALYTIC GEOMETRY

∞

As is the case with many fundamental discoveries, the insight that led to analytic geometry is in retrospect very simple and obvious to all, and to cabdrivers in particular. In taxi terms this insight can be phrased: Every intersection corresponds to a street and an avenue, and every street and avenue correspond to an intersection. A more mathematical rendering states that every point corresponds to an ordered pair of numbers, and every ordered pair of numbers corresponds to a point.

Invented by French mathematician-philosopher René Descartes and independently by his countryman, lawyer Pierre Fermat, in the early seventeenth century, analytic geometry unites algebra and geometry via the above correspondences. The point (3,8), for example, is the point which is 3 units to the right and 8 units up from a fixed point called the origin. The numbers 3 and 8 are called, respectively, the point's x-coordinate and y-coordinate, and indicate a point different from that named by the numbers (8,3), the latter point's position being 8 units to the right and 3 units up from the origin. The origin's coordinates are, naturally enough, (0,0). The importance of the order of the numbers is clear to the cabdriver since (3,8), 3rd Avenue and 8th Street, is not the same as (8,3), 8th Avenue and 3rd Street. The intersection of the primary north-south avenue (the y-axis in mathematical terms) with the primary east-west street (the x-axis) is taken to be the cabdriver's reference point or origin.

Although they are quite different points, both (3,8) and (8,3) are to the right (east of) and above (north of) the origin. By convention, points to the left (west of) or below (south of) the origin require negative coordinates. Thus a point 5 units to the left and 11 units up from the origin has coordinates $(-5,11)$, one 5 units to the left and 11 units down from the origin has coordinates $(-5,-11)$, and one 5 units to the right and 11 down from the origin has coordinates $(5,-11)$. (Don't think of New York as a model since the avenue numbers there increase as you go west.)

So what? There certainly must be more to analytic geometry than this, or universities would automatically award three credits to matriculating cabdrivers. Well, once Descartes and Fermat associated points with ordered pairs of numbers, they further observed, and this is crucial, that algebraic equations corresponded to geometric figures. To cite an easy example, we note that the set of points whose respective x- and y-coordinates make the equation $Y = X$ true constitute a straight line. That is, the points (1,1), (2,2), (3,3) and so on (which satisfy or make true the equation $Y = X$) form a line at a 45-degree angle with the x-axis. Equivalently, the intersections of 1st Avenue and 1st Street, 2nd Avenue and 2nd Street, 3rd Avenue and 3rd Street, and so on constitute a line (or diagonal boulevard) forming a 45-degree angle with the primary east-west reference street.

The graph of the equation $Y = 2X$ is determined in a similar manner but is a line which forms a larger angle with the x-axis. Since the points (1,2), (2,4), (3,6) satisfy the equation (each is such that when its x-coordinate is substituted for X and its y-coordinate is substituted for Y, a true statement results), the intersections of 1st Avenue and 2nd Street, 2nd Avenue and 4th Street, 3rd Avenue and 6th Street, and so on lie along this line. Of course, we needn't restrict ourselves to whole numbers; (1.8, 3.6) also lies on the above line as does $(-2.7, -5.4)$, although neither corresponds to an intersection. Likewise, the equation $Y = 2X + 3$ has a graph passing through the points (0,3), (1,5), (2,7), (3.1,9.2), and so on. By graphing two equations on the same pair of axes, we can determine where they cross. This point is called the equations' simultaneous solution, although, since mathematics is timeless, a better word might be "homolocal."

More complicated equations give rise to more interesting curves (points on which may always be plotted by finding ordered pairs of numbers which make the equation in question true). The graph of the equation $Y = X^2$ is a curve known as a parabola, that of $X^2 + Y^2 = 9$ is a circle, and that of $4X^2 + 9Y^2 = 36$, an ellipse. By exploiting techniques which develop this algebraic/geometric rapprochement, problems in geometry can be rephrased algebraically while algebraic relations can be given a geometric meaning. The unity which results laid the framework for the development of calculus later in the seventeenth century and provided a sort of mathematical lingua franca that is still in use today. (Still in use, I might add, even among the Farsi-, Spanish-, Greek-, Hebrew-, and Russian-speaking cabdrivers in New York. The source for this cabdriver talk dates from an experience I had driving a taxi while a graduate student at the University of Wisconsin in Madison. The dispatcher insisted on using military time and mathematical coordinates to guide the neophyte drivers, many of them foreigners, to their destinations. The problem was that Madison is a small city sandwiched among four lakes, and there simply is no coherent system of rectangular coordinates, just curving diagonals slicing between this lake and that peninsula. The drivers regularly got lost until a dispatcher innocent of analytic geometry took over and directed more conventionally via traffic lights, convenience stores, and gas stations.)

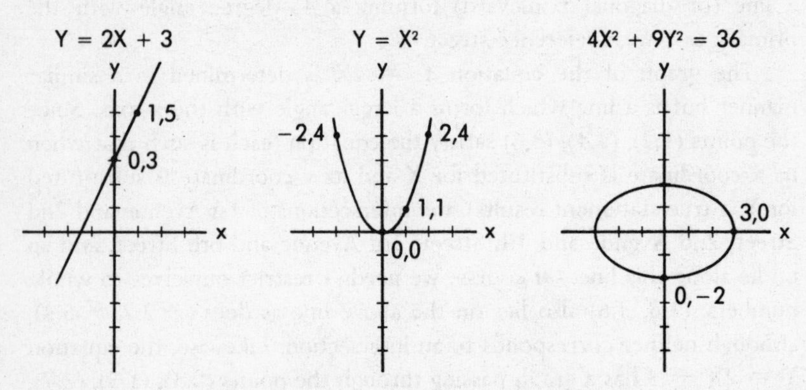

Straight line, parabola, and ellipse

For three dimensions, the generalization is straightforward. Just as any point in the plane can be thought of as an ordered pair of numbers, any point in space can be thought of as an ordered triple of numbers. The point (4,7,5), for example, is 4 units to the east of, 7 units to the north of, and 5 units above some fixed reference point having coordinates (0,0,0); the coordinates simply indicate the distances along the x-axis, y-axis, and z-axis, respectively (rather than just along the first two of these perpendicular axes). Reverting to our cabdriver's perspective, we imagine that (4,7,5) is on the fifth floor of a building at the intersection of 4th Avenue and 7th Street while (4,7, − 1) is in the basement of the same building. The points which satisfy equations in three variables form surfaces in space rather than curves in the plane. The graph of $Z = X^2 + Y^2$, for example, is a paraboloid, a shape something like a rounded coffee cup without a handle, while $X^2 + Y^2 + Z^2 = 25$ has a spherical graph.

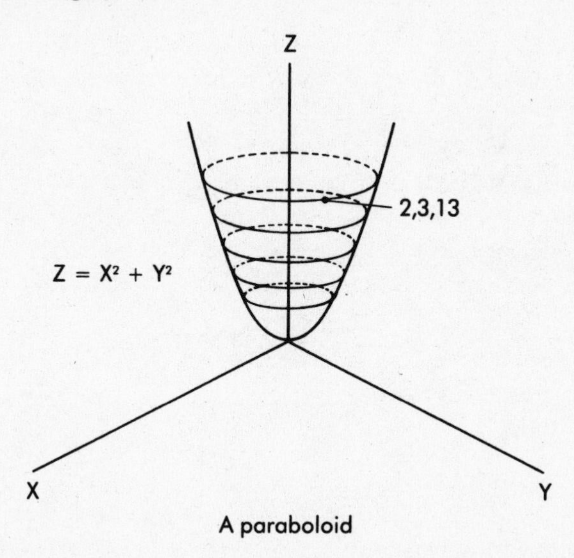

A paraboloid

Generalizations exist for higher-dimensional spaces as well, and the idea of assigning numbers to each of the many dimensions of some entity or other is commonplace in myriad nonmathematical contexts. Thus, the "vector" (4,7, − 1,14) might indicate the aforementioned basement of the building at 4th Avenue and 7th Street at 2 p.m. (1400 hours), while (4,10,9,16) indicates the 9th floor of a building 3 blocks north and 2 hours later.

Like our Hindu-Arabic number system, analytic geometry and its offshoots are so seemingly natural and thus so taken for granted that it sometimes requires a special effort to remember that they're inventions of human beings and not innate aspects of our biological or conceptual makeup. (The great German philosopher Immanuel Kant might disagree with this last remark, but since we're not yet at non-Euclidean geometry, we don't have to pay any attention to him right now.)

ARABIC NUMERALS

A German merchant of the fifteenth century asked an eminent professor where he should send his son for a good business education. The professor responded that German universities would be sufficient to teach the boy addition and subtraction but he would have to go to Italy to learn multiplication and division. Before you smile indulgently, try multiplying or even just adding the Roman numerals CCLXIV, MDCCCIX, DCL, and MLXXXI without first translating them.

Numbers may be eternal and invariant, but numerals, the symbols used to represent numbers, are not, and the above anecdote illustrates how easy it is to take for granted the Hindu-Arabic numerals we use today. The history of numeration systems is a long one extending from prehistoric times to the adoption in the Renaissance of our present system. The heroes of the story are the nameless scribes, accountants, priests, and astronomers who discovered the principles of representing numbers systematically.

These principles—abstract symbolization (as opposed to concrete representations with pebbles, say), positional notation (826 is very different from 628 or 682), a multiplicative base for the system [the numeral 3,243 in our base 10 system, for example, is interpreted as $(3 \times 10^3) + (2 \times 10^2) + (4 \times 10) + (3 \times 1)$, in contrast to its interpretation in a base 5 system, where it would indicate 448: $(3 \times 5^3) + (2 \times 5^2) + (4 \times 5) + (3 \times 1)$], and the holy grail, zero (allowing one to distinguish easily between 36, 306, 360, and 3,006)—are an essential though almost invisible part of our cultural heritage.

[One unconventional "application" of numbers expressed in a different base arises when people celebrate an unwelcome birthday, say their 40th. In a base 12 system, 40 would be written 34: $(3 \times 12) + (4 \times 1)$. With a different base the birthday loses some of its artificial significance. Similar remarks hold with regard to the outbreak of numerological fatuity I expect will be connected to the turn of the millennium in 2000, or 2001 for purists. Changing the base won't always work, however. The senselessness surrounding 666 would probably have attached itself to some other number in a base 5 world where 666 would be innocuously expressed as 10,031 $(1 \times 5^4) + (0 \times 5^3) + (1 \times 5^2) + (3 \times 5^1) + (1 \times 1)$. By an application of the principle of conservation of superstition, 444—124 in our base 10 system—might have acquired mystical significance in such a world.]

Many concrete means of indicating numbers have been used—pebbles of different sizes, knotted skeins of colored threads, the ubiquitous abacus and counting board, and finally the most personal of personal computers: human hands and feet. Counting on our fingers or on our fingers and toes (floppy digits?) is an almost universal phenomenon that ultimately gave rise to the most commonly used written bases. Our base 10 system is certainly a consequence of this, while the French words for 20, 80, and 90—*vingt, quatre-vingts,* and *quatre-vingt-dix*—suggest an older base 20 system. The Mayas, one of the four peoples to invent the principle of positional notation, also used a base 20 system 1,500 years ago to create calendars more accurate than the Gregorian one we use today. Even the ancient Babylonian-Sumerian base 60 system, which survives in our measurement of time, angles, and geographic position, was probably derived from finger counting.

Continuing our sketch, we note that about 2,000 years ago the Chinese invented a written positional numeration system based on powers of 10. About 500 years later the people of southern India independently made the same discovery, but soon thereafter went further and invented zero, a symbol that forever transformed the art of representing and manipulating numbers. Before it received its name, zero was often suggested by an empty space in a numeral or on a counting board. Afterward, the symbol indicated that a space was present or, equivalently, that something was absent. Finally, it became clear that numbers are defined by their properties and that zero was as "propertied" as any other number.

The Chinese borrowed the notion of zero from the Indians, as did the Arabs, who eventually communicated the whole system to Western Europe. The invention of the Hindu-Arabic numeral system can fairly be said to be one of the most important technical discoveries of mankind, ranking with the invention of the wheel, fire, and agriculture.

[By the way, the numbers mentioned in the first paragraph add up to MMMDCCCIV. What's their product?]

AREAS AND VOLUMES

∞

Did you ever read a mammoth generational saga near the end of which a character utters some perfectly vacuous truism which nevertheless resonates deeply with the rest of the book and makes you review everything in its light? My reaction to the formula for the area of a rectangle is somewhat similar. I'm conscious of its obviousness, yet aware as well of its connections to splendid seams of mathematical gold. Let's skip the gold, however, and turn to a more pallid color—manila. A manila envelope lies before me; its dimensions are 10 inches by 13 inches, and thus its area is 130 square inches. Being extraordinarily clever, I didn't count out 130 inch-wide squares to determine this area; all I did was to multiply the envelope's length by its width.

Clear enough, but every other formula for the area of a figure in the plane is a consequence of the simple fact that the area of a rectangle is equal to the length of its base times that of its height; in equation form, $A = BH$. A few important examples follow, a couple of them, like the above, as old as mathematical knowledge itself (pre-Egyptian).

Since a square has a base and height which are equal, the formula for the area of a square is $A = S^2$, S being the length of the square's side. (The corresponding formulas for the perimeters of rectangles and squares are, respectively, $P = 2B + 2H$ and $P = 4S$.)

Although its name may suggest two closely bound letters sent through the mail together, a parallelogram is a four-sided figure—in

Latin English a quadrilateral—whose opposite sides are parallel. Its area can also be found by using the formula, A = BH, where B is again the length of the base or bottom, but H this time is the shortest distance from the top to the bottom (extended if need be), the perpendicular height as opposed to the slant height.

Since any triangle can be considered to be half of a rectangle or parallelogram (a triangular chip off the old rectangular block), the formula for the area of a triangle is simply A = 1/2 × BH, where, as before, B is the base of the triangle and H is the shortest distance from the top vertex to the base (extended if need be).

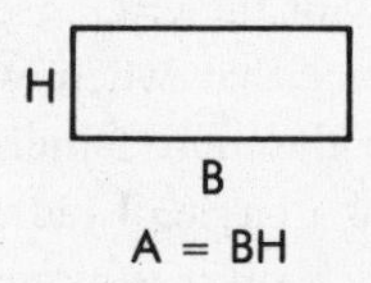

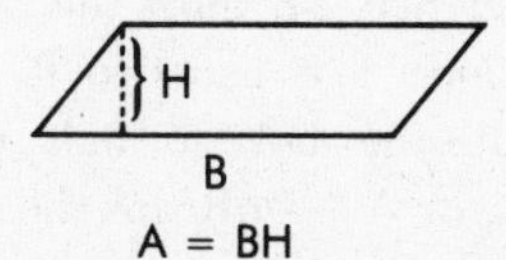

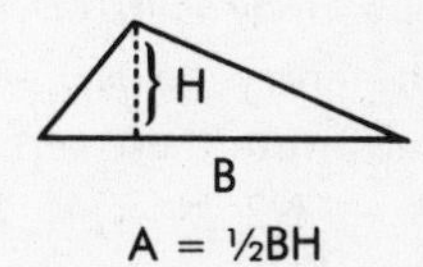

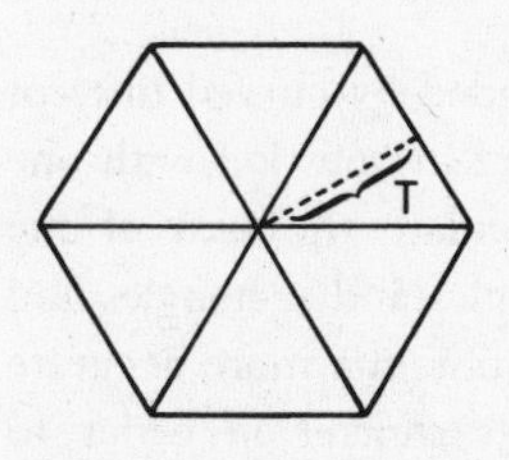

Since the hexagon may be divided into triangles, its area equals ½TP, where T is the distance from its center to one of its sides and P is its perimeter.

But as the number of sides in the inscribed polygon increases, T approaches the circle's radius R and the perimeter P approaches its circumference C.

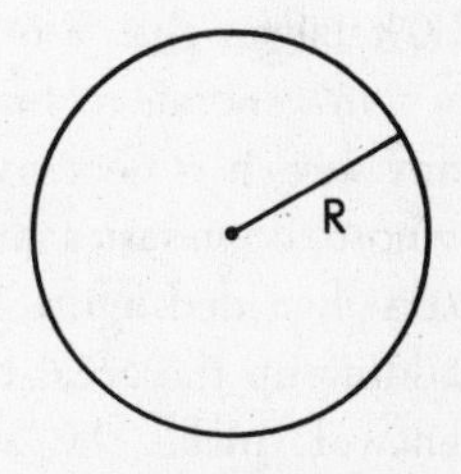

Thus the area of the circle equals ½R(2πR) or, more familiarly, πR^2. $A = \pi R^2$.

Basic area formulas

Whether wallpapering an odd-shaped room or painting the walls of a Gothic cathedral, you may find the area of any plane figure that is bounded by straight lines simply by dividing and conquering: Partition the figure into triangles and rectangles, find the area of each of these triangles and rectangles, and then add up these areas to get the area of the whole figure. Using this method on regular polygons (a polygon such as a triangle, quadrilateral, pentagon, or dodecagon all of whose sides and angles are equal), one can derive the formula $A = 1/2 \times TP$, where P is the perimeter or distance around the regular polygon, and T is the perpendicular distance from the central point of the polygon to a side.

Also, since a circle can be considered to be the limiting figure toward which a sequence of inscribed regular polygons converges, we can use the notions of partitioning and limits to prove that the area of a circle is also given by $A = 1/2 \times TP$, where, as before, T is the distance from the "central point" or center of the circle to a "side" of the circle, and P is the "perimeter" of the circle. Since the perimeter of a circle is its circumference, 2π times the radius R of the circle (see the entry on *pi*), and since T is equal to R, we find that when we substitute these terms into the above formula, we get the more familiar $A = 1/2 \times R \times 2\pi R$, or $A = \pi R^2$ for the area of a circle of radius R (which explains, among other things, why an 8-inch pizza is almost 80% bigger than a 6-inch one).

In general, to find the area of a figure bounded by curved lines of any sort, it is necessary to approximate the figure in question with one whose boundaries are straight lines. Then calculate the area of the straight-sided figure by partitioning it into triangles and rectangles and adding up the areas of these simpler shapes. To obtain a more accurate answer, make the approximating line segments shorter in order to conform more closely to the curved lines, and then partition and add once again. This "method of exhaustion" or successive approximation of curved areas with rectangles and triangles goes back to Archimedes and is the idea behind the definite integral, defined to be the limiting value of these approximating sums, and, incidentally, a concept at the base of much of the applicability of calculus and mathematical analysis.

A similar approach may be taken to the volume of figures in space. The basic building block of such figures is the rectangular box whose volume V is determined by multiplying its length times its width times its height, $V = LWH$; $V = S^3$ for a cube of side S. The volume of a parallelepiped, the six-sided figure whose opposite faces are parallelograms, is given by the same formula but, once again, we interpret width and height to mean the perpendicular distance between opposite sides. Other useful volume formulas are those for a cylindrical can (a phrase as redundant as "new innovation")—$V = \pi R^2 H$, where R is the radius of the can and H is its height; for a cone—$1/3 \times \pi R^2H$, where R is the radius of the base of the cone and H is its perpendicular height; and for a sphere—$4/3 \times \pi R^3$, where R is the sphere's radius.

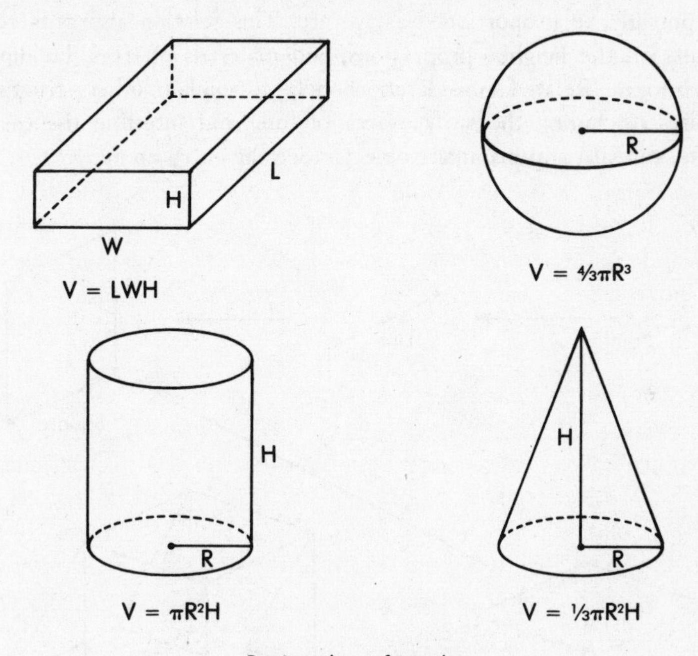

Basic volume formulas

As with area, the exact volume of a more complicated figure may be calculated by finding the limit of successive approximations to the figure with rectangular boxes; i.e., by again using Archimedes' method of exhaustion as formalized in the integral calculus. There are, it should also be mentioned, natural generalizations of area and volume formulas to higher-dimensional hyperspace (hypercubes being the basic building blocks) as well as more theoretical issues concerning the nature and properties of areas and volumes (of surfaces, solids, and arbitrarily chosen sets).

There is an intriguing interplay among the notions of area, volume, and basic physics. Note that the support needed by people, animals, and general structures to stand upright is proportional to their cross-sectional areas, while their weight is proportional to their volumes. For example, quadrupling the height of a structure and preserving its proportions and material makeup will result in a 64-fold (4^3) increase in its weight, but only a 16-fold (4^2) increase in its ability to support the weight. This is why any 25-foot-tall monster ambling about the Himalayas or sunning himself on some beach in the Bermuda Triangle could not possibly be proportioned as we are. This relation also puts constraints on the heights, proportions, and materials of trees, buildings, and bridges. Related considerations help to explain other structural features (including the surface area of lung and intestine tissues) of plants, animals, and inanimate objects. (See the entry on *fractals.*)

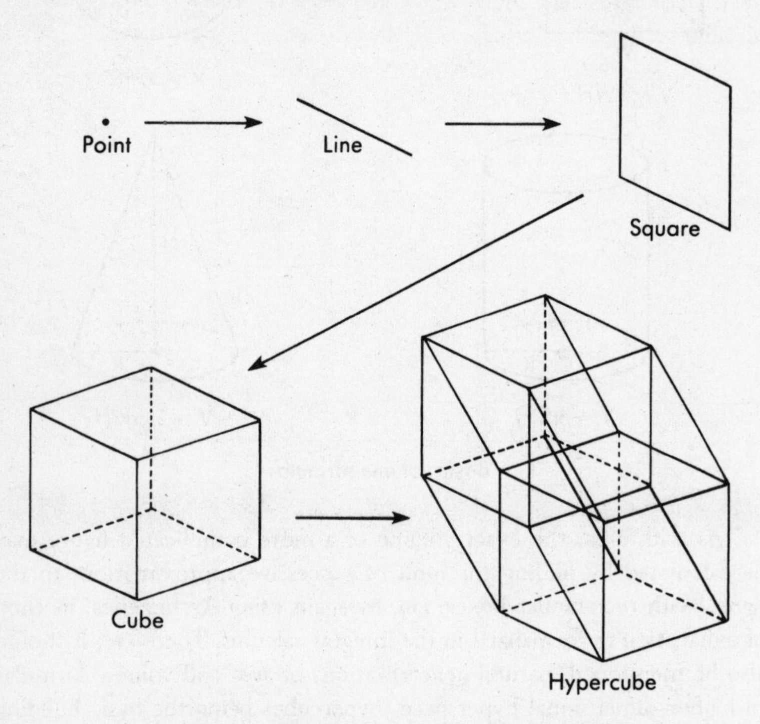

A two-dimensional representation of a hypercube

I repeat what I said at the beginning of this entry. Although the formula A = BH is trivial in one sense (just the archetypal illustration of multiplication), variants, refinements, and applications of it resound throughout mathematics.

Finally, knowledge of these formulas for areas and volume does not always guarantee a gut feel for expanse and voluminosity. One example: The Grand Canyon in Arizona is 217 miles long, ranges in width between 4 and 18 miles, and is up to 1 mile deep. If we conservatively take its average width to be 6 miles and its average depth to be .3 mile, then its volume is 390.6 cubic miles, which, upon multiplication by $5{,}280^3$, yields 5.75×10^{13} cubic feet. If we divide this figure by 5 billion, the human population of the earth, we come up with approximately 11,500 cubic feet of space in the Grand Canyon for every human being on earth. Calculating the cube root of this figure to be about 22.5 feet, I conclude that there would be room in the Grand Tenement for 5 billion cubicle apartments 22.5 feet on a side.

BINARY NUMBERS AND CODES

Binary numbers are ones that consist entirely of 1's and 0's. Although known to ancient Chinese mathematicians, they were first seriously investigated by the German mathematician and philosopher Gottfried Wilhelm von Leibniz, who was motivated by metaphysical considerations of being vs. nonbeing. Because many (all?) phenomena can be reduced to complex sequences of on-off, open-shut, yes-no dichotomies, and because at least computers function in this way, binary numbers and codes have long since descended from the metaphysical realm to the mundane.

In any case, how do we fit our Arabic numerals into this more austere garb of 0's and 1's? Examples serve better than explanation here. The number 53 is expressed as $32 + 16 + 4 + 1$, each of the terms being a power of 2 (1 is thought of as the zeroth power of 2, 2^0). We hence define 110101 to be the binary representation of 53, each 1 or 0 indicating the presence or absence of a power of 2. I.e., $53 = (1 \times 2^5) + (1 \times 2^4) + (0 \times 2^3) + (1 \times 2^2) + (0 \times 2^1) + (1 \times 2^0)$. As with Arabic numerals, the positions of the digits determine their value.

More examples: The number $83 = 64 + 16 + 2 + 1$, so, expressed in binary terms, it is 1010011—$(1 \times 2^6) + (0 \times 2^5) + (1 \times 2^4) + (0 \times 2^3) + (0 \times 2^2) + (1 \times 2^1) + (1 \times 2^0)$. Likewise, $217 = 11011001$. The numbers from 1 to 16 expressed in binary form are: 1, 10, 11, 100, 101, 110, 111, 1000, 1001, 1010, 1011, 1100, 1101,

1110, 1111, 10000. Moving in the other direction, we translate 110100 as 52, 1111100 as 124, and 1000000000 as 512 (2^9). Once Arabic numbers are expressed in binary form, we use the same arithmetic rules and algorithms (e.g., "carrying") to work with them, remembering only that we're dealing with powers of 2 and not powers of 10.

Not limited to numbers alone, binary codes can also be used quite generally and in different ways. For example, if statements are assigned truth values (1 for truth, 0 for falsity), then basic operations of logic—negating a statement, connecting two statements with an "and," "or," or "if . . . , then . . . ," etc.—are easily accomplished by simple syntactical operations or, physically, by simple electronic circuits. (See the entry on *tautologies.*) A statement preceded by a "not" is assigned the truth value 0 or 1 depending on whether the original statement is assigned a 1 or a 0. The statement formed by conjoining two other statements with an "and" is assigned a truth value of 1 only if both the shorter statements are assigned the truth value 1. [These logical operations are called Boolean in honor of the nineteenth-century English mathematician George Boole, who, according to Bertrand Russell's hyperbolic (hyperboolic?) estimate, discovered pure mathematics.]

Coding up letters and other symbols as sequences of 0's and 1's is also no problem, since each character is assigned a different sequence of binary digits or bits. According to the standard ASCII computer conventions, each symbol has an 8-bit code (8 bits = 1 byte), and there are 256 (2^8) such codes in all—one for each of the 52 letters, lowercase and capitalized; for the digits from 0 to 9; and for punctuation, arithmetical, control, and miscellaneous symbols. P has code 01010000; V, 01010110; b, 01100010; t, 01110100; ", 00100010; &, 00100110, and so on. These codes are used for word processing, in which applications the symbols are generally not operated upon as they are in arithmetic but merely listed as text.

To illustrate how a single binary number can encode a large body of information, consider that this entry contains slightly over 5,600 symbols—letters, digits, blank spaces, and punctuation marks. Each of them has an 8-bit code, and so if we concatenate all these codes, we come up with an approximately 45,000-bit sequence that we can take to be the binary representation of the entry. We could do the same thing with the entire book and come up with the book's binary code. Or still more ambitiously, if we arrange all the books in the Library of

Congress alphabetically and chronologically by author and publication date and then concatenate their sequences, we come up with the binary number representing all the information in the Library of Congress.

Exactly what information is (at least one very useful technical definition of it) is answered by the mathematical theory of information. A rich field with many applications to biology, linguistics, and electronics, the theory is couched in the language of bits, each bit of information conveying one binary choice. [Hence 5 bits, for example, convey 5 such choices and are sufficient to distinguish from among 32 (or 2^5) alternatives, there being 32 (2^5) possible yes-no sequences of length 5.] Bits serve too as units in the numerical measure of such notions as the entropy of information sources, the capacity of communication channels, and the redundancy of messages.

From information theory to the dots and dashes of Morse code and the thick and thin lines of supermarket bar codes, binary numbers and codes are now pervasive and commonplace. (The world is breaking into bits and PCs.) I think they are best understood, however, when approached with at least some of Leibniz's original appreciation for their metaphysical primacy. Information, computers, entropy, complexity—all such fundamental terms and ideas stem in part from this most elemental of all codes, 1 or 0, substance or nothing, yin or yang, to be or not to be. It's easy to get carried away by an orgy of synonymous oppositions, so I'll stop and merely note that a universe that was all substance would be indistinguishable from one that was completely empty and thus some binary dichotomy is a precondition for a nontrivial universe as well as for thought itself.

CALCULUS

∞

Calculus, independently discovered in the late seventeenth century by Isaac Newton and Gottfried Wilhelm von Leibniz, is the branch of mathematics that deals with the fundamental concepts of limits and change. Like the axiomatic geometry of the ancient Greeks, it has had a profound effect on the thinking of scientists, mathematicians, and the general public almost from its conception. This is in part due to the power, beauty, and versatility of its ideas and techniques and in part due to its association with Newtonian physics and the metaphor of the universe as a giant clockwork governed by calculus and timeless differential equations. Developments in modern physics have made this metaphor considerably less cogent in recent decades, while computer advances have put into question the once unrivaled position of calculus in the curriculum. Nevertheless, calculus remains one of the most essential branches of mathematics for the scientist and engineer, and increasingly for the economist and businessman as well.

It's useful, although a bit simplistic, to divide the subject into two parts: differential calculus, which deals with rates of change, and integral calculus, which is concerned with summing up quantities that are changing. Beginning with differential calculus as most treatments do, let's assume you leave Philadelphia after lunch and head toward New York on the New Jersey Turnpike. Your car has a clock and an odometer, but no speedometer, and you realize that your speed varies with the traffic, the music, your mood. One question you might naturally ask yourself is how you would determine your instantaneous velocity at a particular time, say one o'clock. Assume you're looking for a theoretical definition more than for a practical method.

A rough approximation to the answer could be obtained by determining your average velocity over the interval between 1:00 and 1:05. Remembering that average velocity is equal to the distance traveled divided by the time required, you could use the odometer to find the distance traveled in these five minutes and then divide this distance by 1/12th of an hour (five minutes). A better approximation would result from finding your average velocity between 1:00 and 1:01. Determine the distance traveled in this one-minute interval and then divide it by 1/60th of an hour. This latter figure would generally be closer to your instantaneous velocity at 1:00 since there is less time for variations in your velocity. A still better approximation to your instantaneous velocity would be obtained by finding your average velocity over the ten-second interval between 1:00:00 and 1:00:10. Again, determine the distance traveled in this ten seconds and divide it by 1/360th of an hour.

This certainly isn't a very efficient method, but it does lead to the theoretical definition of your instantaneous velocity. Your instantaneous velocity at any given instant of time is, by definition, the limit of your average velocities over smaller and smaller time intervals containing the time instant in question. "Limit" is a tricky term here (see the entry on *limits*), but I think the intuition behind it in this case is clear. Furthermore, and this is also important, if the distance you've traveled along the New Jersey Turnpike is given by a formula dependent only on the time you've traveled, calculus provides techniques to enable you to find your instantaneous velocity by manipulating the formula in question. If you were to graph this formula relating the distance traveled (indicated on the y-axis) to the time elapsed (on the x-axis), the velocity at any instant would correspond to how steep the graph was at the given point, in other words to its slope at that point.

The definition and techniques are quite general and arise naturally whenever one is interested in the primeval question: How fast is it changing? As above, we're often interested in how fast a particular quantity changes with respect to time. How fast were we driving at one o'clock? How fast will the oil spill be spreading in three days? How fast was the shadow lengthening one hour ago? Often, however, we're interested in more general rates of change. How fast will our profits grow with respect to the number of widgets manufactured when we're manufacturing 12,000 of them per day? How fast will the temperature of an enclosed gas change with respect to its volume when the volume is 5 liters? How fast will revenues grow with respect to capital investment when the latter is 800 million dollars (other factors holding constant)? As long as the relation between the quantities in question is known, the techniques of differential calculus can be used to determine the rate of change—generally referred to as the "derivative"—of one quantity with respect to the other. [If the relationship between X and Y is given by a formula $Y = f(X)$ (see the entry on *functions*), then the derivative is indicated by a related formula usually symbolized as $f'(X)$; Leibniz's notation for it is dY/dX. The derivative formula tells us how fast Y is changing with respect to X at any point X.]

As in much of mathematics, knowing formulas, in this case derivative formulas obtained by these techniques, is not by itself very valuable. Every calculus student "knows" that the derivative of $Y = X^N$ is NX^{N-1}. To demonstrate the superficiality of this knowledge, I taught my children when they were preschoolers to always respond NX^{N-1} when I asked them what the derivative of X^N was. They too "knew" calculus.

Many kinds of problems do become easily solvable once we possess an understanding of the derivative. Since profits, for example, commonly rise and then fall as a function of the number of widgets manufactured, we know that the rate of change of profit with respect to widgets is first positive (profit increases with the number of widgets made) and then negative (profit decreases if we continue to manufacture more of them). If we know the precise relation between profits and widgets, we can determine, by finding where the derivative or rate of change is zero, how many widgets to manufacture in order to maximize our profits. We can use the same technique more generally to optimize scarce resources.

To get a brief taste for integral calculus now, assume you're back on the New Jersey Turnpike (the royal road to calculus), but this time your car is equipped with a clock (say it reads 2:00) and a speedometer but no odometer. The monotony of driving having put you in a reflective mood, you wonder how, in terms of your velocity, you might derive exactly how far you will have traveled in the next hour. If you maintain a constant velocity of 50 miles per hour, the problem is trivial: You will have traveled exactly 50 miles.

Since your velocity varies considerably, however, you might try to approximate the answer in the following way. Examine your speedometer at 2:02:30 and assume that your speed (say 60 mph) is roughly constant over the interval between 2:00 and 2:05. Since the distance traveled in any given time interval is equal to the product of the velocity and the length of the time interval, multiply 60 mph by 1/12th of an hour to get an approximation of the distance traveled between 2:00 and 2:05 (about 5 miles). Now examine your speedometer at 2:07:30 and, assuming your velocity (say it's 50 mph now) remains roughly constant for the interval between 2:05 and 2:10, multiply 50 mph by 1/12th of an hour to find a rough value for the distance traveled between 2:05 and 2:10. At 2:12:30 you may be encountering heavy traffic and be traveling only 35 mph. Multiply this velocity by 1/12th of an hour to get an estimate for how far you traveled between 2:10 and 2:15. Continuing in this way, add up all these distances to get an approximation of the total distance you traveled during the hour.

There will be considerably less variation in your velocity over one minute than over five minutes, so if you desire a better approximation, use one-minute rather than five-minute intervals and then add up these little bits of distance as before. Alternatively, you could add up the distances traveled in successive ten-second intervals to obtain an even more accurate assessment of the distance you traveled during the hour. The exact distance traveled is defined to be the limit of this procedure, and this limit is called the "definite integral" of velocity. What the sum would be here depends, of course, on your velocity and the exact way in which it varied over the hour-long interval.

As in the case of rates of change, the procedure is quite general and arises whenever one wonders of a changing quantity: How much does it add up to? The total force exerted by a lake against a restraining dam is approximated, for example, by adding the force against the bottom foot-wide stratum of the dam to the force against the foot-wide stratum above it and then to that sum the force against the next stratum and so on until we reach the top of the dam. We need to do this since water pressure and hence the force exerted by water increase with the water's depth. A better approximation of the force is obtained by dividing the dam into inch-wide strata and adding the forces on each of them, while the exact force is obtained by finding the limit of this procedure—the definite integral. Likewise, if we're trying to determine the total revenue resulting from the sale of widgets whose price is continuously declining with the quantity manufactured, we're led to the concept of the definite integral. [The integral of a quantity $Y = f(X)$ is commonly indicated by $\int f(X)dX$, the $\int$ sign being a stylized S symbolizing "sum."]

Much of the usefulness of the definite integral operation comes from the so-called fundamental theorem of calculus which states that this operation and the other basic operation discussed, finding the rate of change or derivative of one quantity with respect to another, are actually inverse operations; they undo each other's effects. The theorem and the techniques growing out of the two definitions provide us with the tools we need to understand continuously changing quantities. Differential equations (equations involving derivatives—see the entry thereon) are a particularly invaluable example of such tools.

These ideas strongly stimulated the development of mathematical analysis; calculus and differential equations became the language of physics; and the world was forever changed. Remember that the next time you're on the New Jersey Turnpike with a defective speedometer or odometer.

CHAOS THEORY

∞

People often charge that a knowledge of mathematics leads to the illusion of certainty and a consequent arrogance. I think this is false, and as one small example in which just the opposite can be the case, I cite chaos theory. The name of the field, like that of its cousin, catastrophe theory, seems particularly apt for a twentieth-century mathematical creation, but forget the name for a bit. All technical fields like to appropriate common terms and twist them into strained parodies of themselves.

Chaos theory does not deal with anarchist treatises or surrealist and dadaist manifestos, but with the behavior of arbitrary nonlinear systems. For our purposes, a system may be thought of as any collection of parts whose interaction is described by rules and/or equations. The U.S. Postal Service, the human circulatory system, the local ecology, and the operating system of the computer I'm presently using are all examples of this loose notion of a system. A nonlinear system is one in which, again quite loosely speaking, the elements are not linked in a linear or proportional manner (as they are, for example, in a bathroom scale or a thermometer); doubling the magnitude of one part will not double that of another, nor is the output proportional to the input. To understand these systems, people manipulate models of them—physical scale-downs, mathematical formulations, and computer simulations. They hope that properties of these more austere models will shed light on the systems in question.

In 1960, while playing with one such computer model of a simple weather system, meteorologist Edward Lorenz discovered something very strange. Inadvertently plugging in numbers that differed by less than one part in a thousand into his model of the weather, he discovered that the resulting weather projections soon diverged farther and farther until they bore no discernible relation to each other. This, as author James Gleick has observed, was the beginning of the mathematical science of chaos theory.

Although Lorenz's nonlinear model was simplistic and his computer equipment primitive, he drew the correct inference from this divergence of computer-simulated weathers: It was caused by the tiny variations in the system's initial conditions. More generally, systems whose development is governed by nonlinear rules and equations can be extremely sensitive to such minuscule changes, often manifesting unforeseeable and "chaotic" behavior as a result. Linear systems, by contrast, are much more robust, small differences in initial conditions leading only to small differences in final outcomes.

The weather, even this simplified model of it, is not susceptible to long-range prediction because it is too sensitive to almost imperceptible changes in the initial conditions, which changes lead to slightly bigger ones a minute later or a foot away, which slightly bigger ones lead to yet more substantial deviations, the whole process cascading over time into a nonrepetitive unpredictability. If one traces the evolution of this simulated weather, one notices that although certain general constraints are satisfied (no blizzards in Kenya, rough seasonal temperature gradients, etc.), specific long-range forecasts, should anyone have the temerity to make them for a year or two hence, are virtually worthless.

Since Lorenz's work, there have been many manifestations of the so-called Butterfly Effect, the sensitive dependence of nonlinear systems on their initial conditions, in disciplines ranging from hydrodynamics (turbulence and fluid flow) to physics (nonlinear oscillators), from biology (heart fibrillations and epilepsy) to economics (price fluctuations). These nonlinear systems demonstrate a surprising complexity that seems to arise even when the systems are defined by quite elementary rules and equations.

Without going into details, let me remark further that the state of these systems at any given time may be assigned a point in a higher-dimensional mathematical "phase space" and that their subsequent evolution then corresponds to a trajectory in this abstract phase space. The trajectories of these systems turn out to be aperiodic and unpredictable, while the graph of all possible trajectories is often incredibly convoluted and, when examined closely, evinces even more intricacy. Still closer inspection of the system's phase-space trajectories reveals yet smaller vortices and complications of the same general kind. In short, the set of all possible trajectories of these nonlinear systems constitutes a fractal (see the entry on *fractals*) and is one of the many sources of mathematician Benoit Mandelbrot's suddenly ubiquitous geometric monsters.

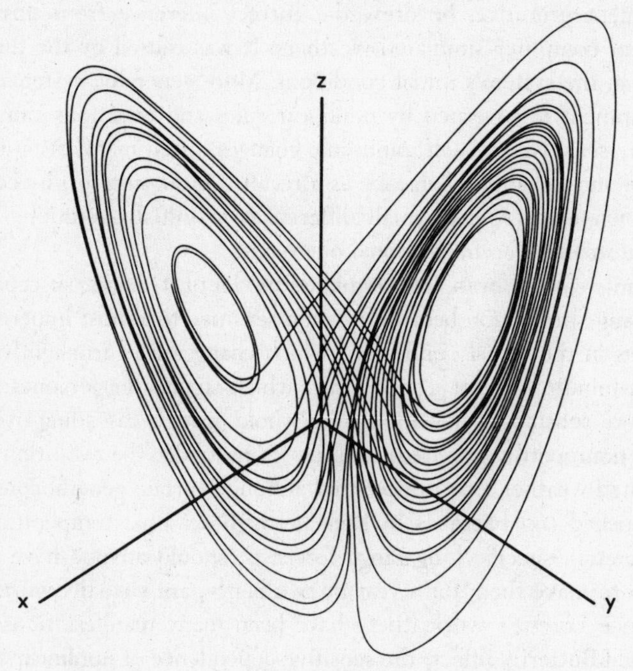

A trajectory in an abstract phase space. The butterfly-shaped figure is called the Lorenz attractor.

I mentioned at the beginning that an appreciation of chaos theory might foster in us a certain caution. One cause for this diffidence is the fact that these systems often behave quite normally and smoothly for a wide range of initial conditions and then suddenly become chaotic when a parameter of the system attains a critical value. Imagine, to cite an intriguingly simple example due to physicist Mitchell Feigenbaum, that the population of a certain animal species is given by the nonlinear formula $X' = RX(1-X)$, where X' is the population one year, X is the population the preceding year, and R is a parameter that varies between 0 and 4. For simplicity we take X and X' to be numbers between 0 and 1, the true population being 1,000,000 times these values.

What happens to the population of this species over time if R = 2? If we assume the population X is now .3 (i.e., 300,000), we know by plugging .3 into the formula that next year the population will be 2(.3)(.7), or .41. To obtain the population the following year, we plug .41 into the formula and find it to be 2(.41)(.59), or .4838. I'll spare you the arithmetic, but the same procedure may be used to find the population for three years from now, for four, five, six years hence, and so on. We find that the population stabilizes at .5. Furthermore, we would find that whatever its original value, the population would still stabilize at .5. This population is termed the steady state population for this value of R.

Similar calculations with a smaller R, say R = 1, demonstrate that whatever its initial value, the population "stabilizes" at 0; it becomes extinct. When R is bigger, say 2.6 to be specific, we find in carrying out these procedures that whatever its initial value the population now stabilizes at approximately .62, the steady state for this R.

So what? Well, let's increase R a little more, this time to 3.2. As before the original population does not matter, but now when we substitute iteratively into the formula $3.2X(1-X)$, we find that the population of the species doesn't stabilize at one value, but eventually settles into a repetitive alternation between two values—approximately .5 and .8. That is, one year the population is .5 and the next it's .8. Let's raise our parameter R to 3.5 and see what happens. The initial population is once again immaterial, but this time in the long run we have the population alternating regularly among four different values, approximately .38, .83, .50, and .88, in successive years. If we again increase R slightly, the population settles down to a regular alternation among eight different values. Smaller and smaller increases in R lead to more doublings of the number of values.

Then suddenly, at approximately R = 3.57, the number of values grows to infinity and the population of the species varies randomly from year to year. [This is a queer sort of randomness, however, since it results from iterating the formula 3.57X(1 − X), and the sequence of populations is thus quite determined by the initial population.] Even stranger than this chaotic variation in the species' population is the fact that a slight increase in R again results in a regular alternation of the species' population from year to year, and a further increase again leads to chaotic variation. These orderly alternations followed by random chaos followed again by windows of regularity are critically dependent on the parameter R, which seems to be a gross measure of the volatility of the model.

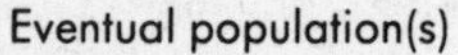

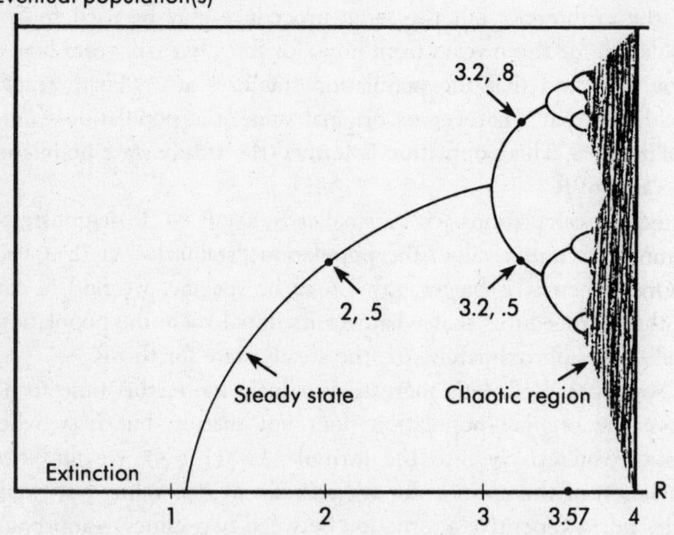

Relationship between values of R, population sizes, and chaos. Chaotic region is very intricate. This type of diagram is due to biologist Robert May.

Although a close-up view of the chaotic region reveals an unexpected intricacy, the annual variation of the population doesn't lend itself to an especially beautiful fractal the way some nonlinear systems do. Nevertheless, there is sufficient complexity so that the following lesson seems clear. If a system as trivial as this single nonlinear equation can demonstrate such chaotic unpredictability, perhaps we shouldn't be quite as assertive and dogmatic about the predicted efforts of various social, economic, and ecological policies on the gigantic nonlinear systems that are the U.S.A. and Planet Earth. The policies' consequences, one would think, are considerably more opaque than are those of the value of R in this model.

It's always dangerous and often idiotic to apply technical results outside their original domain. Thus, I don't intend to suggest by my cautionary remarks above that we should adopt a handwringing passivity simply because we can't always be certain of the effects of our actions; I do mean that chaos theory (and much else) counsels that skepticism and a certain humility should accompany the adoption of any political, economic, or military policy. (For want of a nail, the battle was lost and all that.) A personal anecdote in this regard: I was asked many times during interviews set up to promote my book *Innumeracy* what global pronouncements I could make about the decline of the United States. I must admit I succumbed a few times and ventured far beyond mathematics education to go through the usual dreary litany about the sad state of our society. For the most part, however, I resisted, stating that I could not have predicted the success of *Innumeracy,* much less the demise of American civilization.

One might conclude that these chaotic, nonlinear systems are rare animals, curiosities that bedevil only mathematicians and various fringe scientists. That this is most emphatically not the case was vividly expressed by mathematician Stanislaw Ulam, who commented that calling chaos theory "nonlinear science" is like calling zoology "the study of non-elephant animals." Now that more tools for dealing with nonlinearity have been developed, scientists are realizing that it's no longer necessary to screen out noise, static, and friction or to ignore turbulence, arrhythmia, and "random" contingency. These phenomena, like fractals, are everywhere, and it is their linear cousins that are, like elephants, rare.

COINCIDENCES

∞

Coincidences fascinate us. They seem to compel a search for their significance. More often than some people realize, however, they're to be expected and require no special explanation. Surely no cosmic conclusions may be drawn from the fact that I recently and quite by accident met someone in Seattle whose father had played on the same Chicago high school baseball team as my father had and whose daughter is the same age and has the same name as my daughter. As improbable as this *particular* event was (or as particular events always are), that *some* event of this vaguely characterized sort should occasionally occur is very likely.

More precisely, it can be shown, for example, that if two strangers sit next to each other on an airplane, more than 99 times out of 100 they will be linked in some way by two or fewer intermediates. (The linkage with my father's classmate was more striking. It was via only one intermediate, my father, and contained other elements.) Maybe, for example, the cousin of one of the passengers will know the other's dentist. Most of the time people won't discover these links, since in casual conversation they don't usually run through all their 1,500 or so acquaintances as well as all their acquaintances' acquaintances. (I suppose with laptop computers becoming more popular they could compare their own personal databases and even those of people they know. Perhaps exchanging databases might soon be as common as leaving a business card. Electronic networking. Hellacious.)

There is a tendency, however, to home in on likely co-acquaintances. Such connections are thus discovered frequently enough so that the squeals of amazement that commonly accompany their discovery are unwarranted. Similarly unimpressive is the "prophetic" dream which traditionally comes to light after some natural disaster has occurred. Given the half billion hours of dreaming each night in this country—2 hours per night for 250 million people—we should expect as much.

Or consider the famous birthday problem in probability theory. One must gather together 367 people (one more than the number of days in a leap year) in order to ensure that 2 of them share a birthday. But if one is willing to settle for a 50-50 chance of this happening, only 23 people need be gathered. Rephrasing, I note that if we imagine a school with thousands of classrooms each of which contains 23 students, then approximately half of these classrooms will contain 2 students who share a birthday. No time should be wasted trying to explain the meaning of these or other coincidences of similar type. They just happen.

One somewhat different example concerns the publisher of a stock newsletter who sends out 64,000 letters extolling his state-of-the-art database, his inside contacts, and his sophisticated econometric models. In 32,000 of these letters he predicts a rise in some stock index for the following week, say, and in 32,000 of them he predicts a decline. Whatever happens, he sends a follow-up letter but only to those 32,000 to whom he's made a correct "prediction." To 16,000 of them he predicts a rise for the next week, and to 16,000 a decline. Again, whatever happens, he will have sent 2 consecutive correct predictions to 16,000 people. Iterating this procedure of focusing exclusively on the winnowed list of people who have received only correct predictions, he can create the illusion in them that he knows what he's talking about. After all, the 1,000 or so remaining people who have received 6 straight correct predictions (by coincidence) have a good reason to cough up the $1,000 the newsletter publisher requests: They want to continue to receive these "oracular" pronouncements.

I repeat that a useful distinction in discussing these and other coincidences is that between generic sorts of events and particular events. Many situations are such that the particular event that occurs is guaranteed to be rare—a certain individual winning the lottery or a specific bridge hand being dealt—while the generic outcome—someone's winning the lottery or some bridge hand being dealt—is unremarkable. Consider the birthday problem again. If all that we require is that 2 people have some birthday in common rather than any particular birthday, then 23 people suffice to make this happen with a probability of 1/2. By contrast, 253 people are needed in order for the probability to be 1/2 that one of them has a specific birth date, say July 4. Particular events specified beforehand are, of course, quite difficult to forecast, so it's not surprising that predictions by televangelists, quack doctors, and others are usually vague and amorphous (that is, until the events in question have occurred, at which time the prognosticators like to assert that these precise outcomes were indeed foreseen).

This brings me to the so-called Jeane Dixon effect, whereby the few correct predictions (by psychics, disreputable stock newsletters, whomever) are widely heralded and the 9,839 or so false predictions made annually are conveniently ignored. The phenomenon is quite widespread and contributes to the tendency we all have to read more significance into coincidences than is usually justified. We forget all the premonitions of disaster we've had which didn't predict the future and remember vividly those few which seemed to do so. Instances of seemingly telepathic thought are reported to everyone we know; the incomparably vaster number of times this doesn't occur are too banal to mention.

Even our biology conspires to make coincidences appear more meaningful than they usually are. Since the natural world of rocks, plants, and rivers doesn't seem to offer much evidence for superfluous coincidences, primitive man had to be very sensitive to every conceivable anomaly and improbability as he slowly developed science and its progenitor, "common sense." Coincidences, after all, *are* sometimes quite significant. In our complicated and largely man-made modern world, however, the plethora of connections among us appears to have overstimulated many people's inborn tendency to note coincidence and improbability and led them to postulate causes and forces where there are none. People know more names (not only family members' but also those of colleagues and a myriad of public figures), dates (from news stories to personal appointments and schedules), addresses (whether actual physical ones or telephone numbers, office numbers, and so on), and organizations and acronyms (from the FBI to the IMF, from AIDS to ASEAN) than ever before. Thus, although it is a very difficult quantity to measure, the rate at which coincidences occur has probably risen over the last century or two. Still, for most of them it generally makes little sense to demand an explanation.

In reality, the most astonishingly incredible coincidence imaginable would be the complete absence of all coincidences.

[Brief derivations of birthday statements (see also the entry on *probability*): (1) The probability of 2 people having different birthdays is 364/365; of 3 people having different birthdays (364/365 × 363/365); of 4 people (364/365 × 363/365 × 362/365); of 23 (364/365 × 363/365 × 362/365 × . . . × 344/365 × 343/365), which product turns out to equal 1/2. Thus the complementary probability that at least 2 people share a birthday is also 1/2 (1 minus the above product). (2) The probability that someone does not have a July 4 birthday is 364/365; the probability that neither of 2 people has a July 4 birthday is $(364/365)^2$; none of 3 $(364/365)^3$, none of 253 $(364/365)^{253}$, which turns out to equal 1/2. Thus the complementary probability that at least one of the 253 people has a July 4 birthday is also 1/2, $1 - (364/365)^{253}$.]

COMBINATORICS, GRAPHS, AND MAPS

∞

You are a member of that dying breed of craftsmen known as map colorers and you have before you a plane map of the Reticulated Provinces of Convolutia. The university publishing house which employs you has fallen on hard times and you wonder if you will be able to color the map with at most four colors and ensure, of course, that adjacent countries which share at least part of a border are colored differently. The four-color theorem guarantees that you will be able to accomplish your avowed task no matter how many countries there are and no matter how the countries are arranged so long as they are connected regions (that is, you can't have a bit of Schizostan over here and another bit a thousand miles away).

The conjecture of the sufficiency of four colors for a flat map was stated in the middle of the nineteenth century, but despite much attention it remained unresolved until 1976, when the American mathematicians Kenneth Appel and Wolfgang Haken proved that four colors were indeed enough. Earlier efforts had shown that at most seven colors were needed to color maps on the surface of a torus (a figure having the shape of an inner tube), but, not being accustomed to drawing maps on doughnuts, most people find this proposition lacking the natural appeal of the four-color theorem.

The proof of the four-color theorem requires the use of a computer to check all the myriad possibilities associated with various types of map configurations. This is a novel development, on the surface at odds with the very notion of a mathematical proof. Proofs, it has been tacitly assumed since the time of the Greeks, should be capable of being grasped and verified by humans. Moreover, they should be logically convincing. (See the entry on *QED.*) A proof like that of the four-color theorem which requires such extensive use of computers is not graspable or verifiable in the same way as other mathematical demonstrations. Neither is it logically hermetic. Although the probability of a mistake on any given configuration is minuscule, the number of permutations and arrangements checked is so vast that we may conclude only that the theorem is probably true. But "probably true" is not the same as "conclusively proved."

Some mathematicians have noted that there are a few theorems in mathematical group theory so complicated and voluminous that some of the same objections may be made about them even though they make no use of computers. However one views these matters, it's clear that the proof of the four-color theorem is not elegant, compelling, or natural. It certainly doesn't belong in what mathematician Paul Erdos calls God's Book, a collection of ideal proofs which do have these properties. Still, it is an impressive resolution of an old problem.

Unfortunately, the mathematical fallout from the result has not been as impressive as that from another old puzzle, Leonhard Euler's Königsberg bridge problem. Euler began his classic 1736 paper (which many take as having inaugurated the field of combinatorics) by discussing the layout of the city of Königsberg in East Prussia. The city is located on the banks and on two islands of the river Pregel. The various parts of the city were joined by seven bridges, and on Sundays people would stroll about the town. The question arose whether it would be possible for residents to begin their walk at home and to return there after traversing each river bridge exactly once. Euler showed that such a route did not exist. His basic insight was that any such route would have to enter each part of the city as many times as it departs from it and thus would require an even number of bridges into each part of the city, a condition not fulfilled by Königsberg.

Euler represented the different parts of the city as points and the bridges between them as lines. The resulting collection of points and connecting lines (or vertices and edges) constituted a graph, and in the remainder of his paper Euler studied the general problem: In which graphs of this sort is it possible to find a path running through each of the lines just once before returning to the starting point? (Note that in contrast to the Königsberg situation, it is possible to find such a path for a Star of David graph, drawn by superimposing two triangles, one pointing upward, the other a little lower and pointing downward. Note also that every vertex of this latter graph has an even number of edges associated with it.)

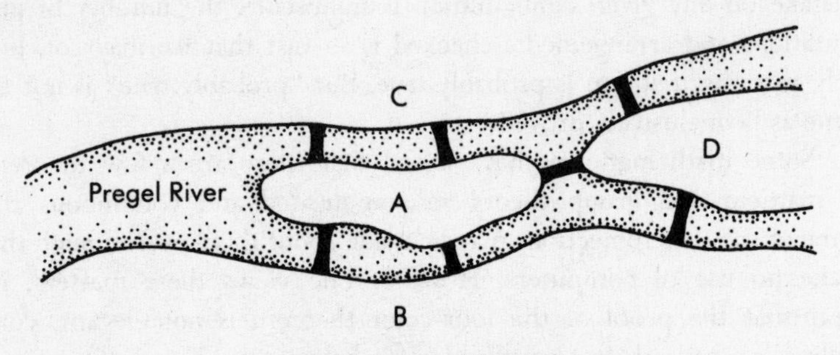

The seven bridges of Königsberg

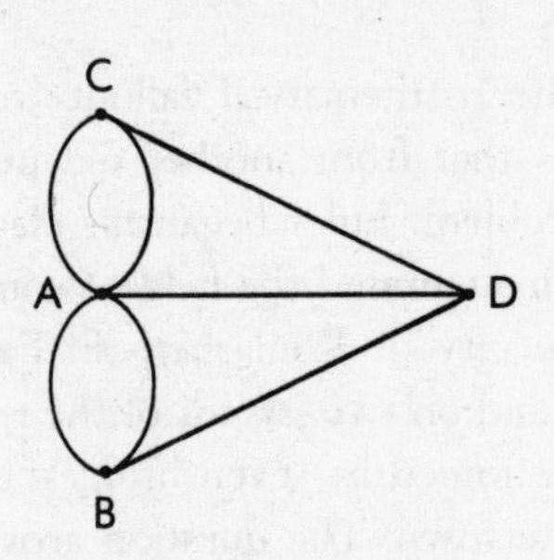

Euler's graph of the city

This seemingly simpleminded way of representing mathematical relations with points and connecting lines is now an indispensable tool in graph theory. It may be used, for example, to represent the network of acquaintanceship among a group of people (lines joining acquaintances), the tree of all possible outcomes in a complicated sequence of choices, round-robin matchings in a sports tournament, connections in an integrated circuit, and much else. The four-color theorem, by contrast, seems to be a dead end.

It's impossible to predict whether a problem will lead to further developments and insights or to a blind alley. (It's irrelevant but interesting that "cul-de-sac," "blind alley," "no exit," and "dead end" all connote darkness, in this case the absence of mathematical alternatives and vision.) Consider the following two puzzles and decide which one is the cul-de-sac.

Start with any positive integer and, if it's even, divide it by 2, but if it's odd, multiply it by 3 and then add 1. Follow the same rules on the resulting integer and iterate this procedure. The sequence generated from 11 is: 11, 34, 17, 52, 26, 13, 40, 20, 10, 5, 16, 8, 4, 2, 1, 4, 2, 1, . . . , while that generated from 92 is: 92, 46, 23, 70, 35, 106, 53, 160, 80, 40, 20, 10, 5, 16, 8, 4, 2, 1, 4, 2, 1, ... The question is whether every positive number eventually reaches the 4-2-1 cycle and the belief is that this is so, but it's never been proved.

The other problem may be phrased in terms of guests at a small dinner party. The question is: What is the smallest number of guests who need be present so that it is certain that at least 3 of them will know each other or at least 3 of them won't? (Assume that if Martha knows George, then George knows Martha.) That the answer is 6 may be seen by imagining that you're a guest at the party. Since you know or don't know each of the other 5, you will either know at least 3 of them or not know at least 3 of them. Suppose that you know 3 of them (the argument is parallel if there are at least 3 you don't know) and consider what relationships hold among your 3 acquaintances. If any 2 of them know each other, then they and you constitute a group of 3 guests who know each other. On the other hand, if none of your 3 acquaintances know each other, they constitute a group of 3 who don't know each other. Thus 6 are sufficient. To see that 5 guests are not enough, imagine yourself at such a smaller party where you know exactly 2 of the other 4 guests, and each of them knows a different one of the 2 people you don't know.

The meta-level question is: Which puzzle leads to something and which doesn't? The answer is that although both are useless in a practical sense, the first is a dead end (at least so far), while the second leads to a whole new branch of combinatorics known as Ramsey theory. Named after English mathematician Frank Ramsey, the theory is concerned with finding the minimum number of elements which satisfy various simple combinatoric conditions. It's full of problems which are maddeningly easy to state and yet remain unsolved even by brute-force computer methods. One such is the generalization to the above which asks for the minimum number of party guests necessary to ensure that there are at least 5 guests who know each other or at least 5 who do not.

P ———— Q – P knows Q.
R ∿∿∿∿∿∿ S – R doesn't know S.

A B C D E

There is no set of 3 people who know each other. Neither is there a set of 3 people who are unacquainted.

A concluding exercise. The four-color theorem states that at most four colors are needed to color any plane map. There are several essentially different small maps which demonstrate that at least four colors are required. Sketch one.

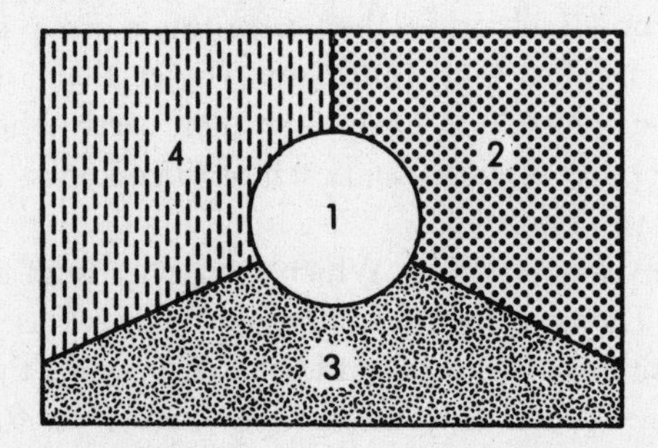

A map requiring four colors. Four colors are always sufficient.

COMPLEXITY OF PROGRAMS

I once knew someone who had read and apparently misunderstood a book on mnemonic devices. To memorize a telephone number, for example, she might have recalled that her best friend had 2 children, her dentist had 5, her college roommate 3, her neighbor on one side had 3 dogs, the one on the other side 7 cats, her older brother had 8 children if you counted those of his wives, and she herself was one of 4 children. The telephone number must be 253-3784. Her algorithms (recipes, programs) were convoluted, inventive, amusing, and always very much longer than what they were designed to help her remember. Sometimes, of course, when such algorithms are intimately linked with a list or a story that one already knows very well, their length is only apparent and their use is reasonable and justifiable. This was not the case with my friend, however, who invariably forgot crucial elements of them.

Assuming you had some interest in doing so, how would you describe the following sequences to an acquaintance who couldn't see them?

(1) 0 0 1 0 0 1 0 0 1 0 0 1 0 0 1 0 0 1 0 0 1 0 0 1 0 . . .
(2) 0 1 0 1 1 0 1 0 1 0 1 0 1 0 1 0 1 1 0 1 0 1 0 1 1 . . .
(3) 1 0 0 0 1 0 1 1 0 1 1 0 1 1 0 0 0 1 0 1 0 1 1 0 0 . . .

Clearly the first sequence is the simplest, being merely a repetition of two 0's and a 1. The second sequence has some regularity to it—a single 0 alternating sometimes with a 1, sometimes with two 1's—while the third sequence is the most difficult to describe since it doesn't seem to evince any pattern at all. Observe that the precise meaning of . . . in the first sequence is clear; it is less so in the second sequence, and not at all clear in the third. Despite this, let's assume that these sequences are each a trillion bits long (a bit is a 0 or a 1) and continue on "in the same way."

Mindful of these examples, we can follow the computer scientist Gregory Chaitin and the Russian mathematician A. N. Kolmogorov and define the complexity of a sequence of 0's and 1's to be the length of the shortest computer program which will generate (i.e., print out) the sequence in question.

What does this mean? Note that a program which prints out the first sequence above can consist simply of the following recipe: Print two 0's, then a 1, and repeat a third of a trillion times. Such a program is quite short, especially compared with the length of the trillion-bit sequence it generates. The complexity of this first sequence may be only about 1,000 bits, or however long the shortest program generating it turns out to be. This depends to some extent on the computer language used to write the program, but no matter what this language is, translating "Print two 0's, then a 1, and repeat" into it isn't going to result in a very long program. (For uniformity's sake, we might as well assume that our programs are written in a very spare machine language of 0's and 1's so that the programs which generate the sequences can themselves be considered to be sequences of 0's and 1's.)

A program that generates the second sequence would be a translation of the following: Print a 0 followed by either a single 1 or two 1's, the pattern of the intervening 1's being one, two, one, one, one, one, one, two, and so on. If this pattern persists, any program which prints out the sequence would have to be quite long so as to fully specify the "and so on" pattern of the intervening 1's. Nevertheless, because of the regular alternation of 0's and 1's, the shortest such program will be considerably shorter than the trillion-bit sequence it generates. Thus the complexity of this second sequence might be only half a trillion bits, or whatever the length of the shortest program generating it is.

With the third sequence (by far the commonest sort) the situation is different. The sequence, let us assume, remains so disorderly throughout its trillion-bit length that no program we might use to generate it would be any shorter than the sequence itself. All any program can do in this case is to dumbly list the bits in the sequence: Print 1, then 0, then 0, then 0, then 1, then 0, then 1, ... There is no way the . . . can be compressed or the program shortened. Such a program will be at least as long as the sequence it's supposed to print out, and thus the third sequence has a complexity of approximately a trillion.

A sequence like the third one which requires a program as long as itself to be generated is said to be random. Random sequences evince no pattern, regularity, or order, and the programs which print them out can do nothing more than direct that they be copied: Print 1 0 0 0 1 0 1 1 0 1 1 . . . . These programs cannot be condensed or abbreviated. The complexity of the sequences they generate is equal to the length of these sequences. By contrast, ordered, regular sequences like the first can be generated by very short programs and have complexity much less than their length.

In some ways sequences like the second are the most interesting, however, since, like living things, they display elements of order and randomness. Their complexity is less than their length, but not so small as to be completely ordered or so large as to be random. The first sequence above may perhaps be compared to a diamond or a quartz crystal in its regularity, while the third is comparable to a cloud of gas molecules or a succession of coin flips in its randomness. The analogue of the second might be something like a rose (or, less poetically, an artichoke), which manifests both order and randomness among its parts.

These comparisons are more than mere metaphors. The reason is that most phenomena can be described via some code, and any such code, whether it be the molecular language of amino acids and DNA molecules or the English language of letters and books, can be digitalized and reduced to sequences of 0's and 1's. (See also the entries on *music, art, and digitalization* and *Turing's test.*) Both DNA and a romance novel are, expressed in their respective codes, sequences like the second one above, evincing order and redundancy as well as complexity and randomness. Similarly, complex melodies lie between simple repetitive beats and formless static (analogous, respectively, to sequences like the first and the third).

The whole of science might even be conceived of in this way. Ray

J. Solomonoff and others have theorized that a scientist's observations could be coded up into a sequence of 0's and 1's. The goal of science then would be to find short programs capable of generating (deriving, predicting) these observations. Such a program, so the story goes, would be a scientific theory, and the shorter it was, relative to the observational phenomena it predicted, the more powerful a theory it would be. Random events would not be predictable except in a very Pickwickian sense by a program which simply listed them. Such a conception of science is quite simplistic and only begins to make sense relative to a well-defined and already fixed scientific framework. Nevertheless, it is suggestive of the generality of the notion of complexity.

The notion of complexity I've defined here is called algorithmic complexity because it gives one the length of the shortest program (algorithm, recipe) needed to generate a given sequence. A related notion is the computational complexity of a sequence. It is defined to be the shortest time necessary for a program to generate the sequence in question. Often in practical problems, time is the more important factor in choosing a program. A short program may take eons to generate the sequence (or, in the Solomonoff conception of science, the prediction) that you want when a longer one will do the job in short order.

Most computational problems (which can, remember, always be coded up into sequences, as can their solutions) take longer to solve the bigger they are. Thus, the so-called traveling salesman problem, which requires that we find the shortest route wending once through each of a number of cities and then returning to the original city, takes longer to solve the more cities we must consider.

An important distinction to make is between those mathematical problems whose computational complexity is an exponential function of their size and those whose computational complexity is a polynomial function of their size. To illustrate in the case of the traveling salesman problem, the size of the problem is determined by N, the number of cities to be visited, and it's unknown whether the time necessary to solve the problem grows exponentially as 2^N or polynomially as N^T, where T is some (generally small) power. (See the entry on *exponential growth.*) A numerical example is telling. Taking $N = 20$ and $T = 3$, we find that $2^N = 2^{20} = 1{,}048{,}576$, whereas $N^T = 20^3 =$ (only) 8,000.

If, as suspected, the computational complexity of the traveling salesman problem grows as 2^N, then the time needed to solve it increases so rapidly that the problem is, for all practical purposes, unsolvable for very large N. (Even parallel processors, computers whose internal architecture is such that they can handle many operations simultaneously —in parallel as opposed to present-day computers' one-step-at-a-time serial processors—won't change this fact.) If the computational complexity grows polynomially as some power of N, then complete solutions can generally be obtained in a reasonable amount of time even for large N.

The close conceptual alliance between notions of economy and constraint and the ideas of algorithmic and computational complexity partially explains the increasing importance of these ideas. Computers of every design have limited time and storage capacity, and more efficient algorithms must be constructed for many problems (airline scheduling and call routing, for example). These concepts of complexity are also, as I've suggested, of theoretical and philosophical significance. Many theorems, including Gödel's incompleteness theorem (see the entry on *Gödel*), have natural proofs when expressed in terms of complexity. Other statements, including the so-called P-NP problem (which asks if those problems whose proposed solutions can be checked quickly [polynomially] are such that their solutions may also be discovered quickly), only make sense in the context of complexity theory.

COMPUTATION AND ROTE

∞

"Ambition, Distraction, Uglification, and Derision" is how Lewis Carroll referred to the basic arithmetic operations of addition, subtraction, multiplication, and division. This is how many people, myself included, still view school computation (except for "ambition," which never seemed to belong on the list, "addiction" seeming more appropriate). The reason for this repugnance, I think, is that computation is boring, tiresome, and oppressive. Worse than this, it often forever colors (or should I say discolors) people's views of real mathematics.

Imagine that 90 percent of every course in English up until college was devoted to grammar and the diagramming of sentences. Would graduates have any feeling for literature? Or consider a conservatory which devoted 90 percent of its efforts to the practicing of the scales. Would its students develop an appreciation or understanding of music? The answer, of course, is no, but that, given proper allowance for hyperbole, is what frequently happens in our mathematics classes. Mathematics is identified with a rote recitation of facts and a blind carrying out of procedures. Decades later this robotic mode of behavior kicks in whenever a mathematical topic arises. Countless people feel that if the answer or at least a recipe for finding it doesn't come to them immediately, they'll never get it. The idea of thinking about a problem or discussing it a bit with someone else seems novel to them. Think about a math problem? Discuss it? (See also entries on *substitutability* and *humor.*)

In my opinion, schools' focus on computation is excessive and obsessive. There is nothing wrong, of course, with knowing the addition and multiplication tables and the basic algorithms for dealing with fractions, percentages, etc. In fact, these skills are absolutely essential, even today when a $5 calculator (a part, I would hope, of every kid's school wardrobe) can do all the computing most people will ever need. It's just that after *some* routine drill, these skills should be regarded as tools that further understanding, not as a substitute for understanding.

One unappreciated consequence of this seeming truism is that mathematics should be understood to be continuous with language and logic (see entries on *variables, quantifiers,* and *a mathematical approach*) and not as some isolated collection of mental isometric exercises. In elementary school, for example, there should be units devoted to deciding which arithmetic operation or sequence of operations is called for in a given problem; to estimating very large or very small magnitudes; to mathematically flavored detective stories; to numerical patterns and mechanical puzzles (e.g., Rubik's cube); to board games (such as Monopoly) in which chance is a factor; to mathematical aspects of news stories and sports events (batting averages); and to a myriad of other topics that can be related to a child's life.

If this natural connection between mathematics and ordinary language and thought is established early, then the tables, formulas, and algorithms that come later are justified: They're just a shorthand means to get to the solution. So-called word problems (not a natural ontological class—just any of an infinite variety of mathematical problems expressed in words) would not be aberrations in math classes, but their primary core. The silly plaint "I can do the math but not the word problems" would be heard less frequently in high schools and colleges. Whenever I hear it now, I wonder what the person thinks is the point of the "math." Are pages of polynomials to factor or pages of functions to differentiate their own rationale?

A constant emphasis on computation in early schooling brings with it the tyranny of the right answer, another impediment to learning mathematics and another aspect of the still too common priestly/militaristic model of mathematics instruction. This is the Truth; now do 400 identical problems. In most other fields there is a clear distinction among wrong answers, but too many people believe that if an answer in mathematics is not right, it's wrong—period. Just the opposite is the case. If two people were to add 2/5 and 3/11 and one came up with 5/16 and the other with 39/55, it would be fairly clear that the first one didn't know much about fractions and that the second had only been careless. (Actually, even the first "addition"—$2/5 + 3/11 = 5/16$—could be defended. Perhaps the 2/5 means that a baseball player obtained 2 hits in 5 at-bats and the 3/11 means he got 3 more hits in 11 subsequent at-bats. He would then be 5 for 16, so the above "addition" is justified.)

In arithmetic, algebra, and elementary probability there are usually several ways to solve a problem, and in less well-defined, more difficult problems there are more. (I usually give full credit for wrong answers if only the "math" is wrong but the conception is correct, and partial credit if the approach is a reasonable one.) The popular belief that all wrong answers are equivalent, or even that all right answers are equivalent, lessens the need to think critically, and this accounts for its prevalence among students and, sad to say, among many teachers as well.

The handwringing about how our kids can't do simple computation strikes me as similar to debates that might have raged in fifteenth-century Italy over how students were having trouble with the division algorithms for Roman numerals. It was gradually becoming apparent that a facility for this particular skill was difficult to acquire and, given the new Arabic numeral "software," less useful than it had been. In an attenuated way, this is the situation today. The ability to compute by hand is less important than it used to be, and this is another reason that our fundamentalist focus on computational skills should be discarded.

[Before I went to graduate school, I had a brief sojourn with the Peace Corps and taught mathematics at a secondary school in a very poor district of western Kenya. Despite the absence of materials, the mathematical performance of the students was considerably better than that of students at the richer Nairobi schools where there were plenty of old-fashioned workbooks with page after page of dreary exercises. The emphasis of the small staff at my school was on practical problems (at least at first) and on conceptual understanding. Certainly not a conclusive study, the experience nevertheless intensified my dissatisfaction with traditional pedagogy.]

Mathematics is no more computation than writing is typing. Almost everyone eventually learns how to compute, but as reports on our mathematics education clearly indicate, other higher-level abilities are not being fostered in our children. Many high school students can't interpret graphs, don't understand statistical notions, are unable to model situations mathematically, seldom estimate or compare magnitudes, never prove theorems, and, most distressing of all, hardly ever develop a critical, skeptical attitude toward numeric, spatial, and quantitative data or conclusions. The public and private costs of this innumeracy and general mathematical incapacity are incalculable.

I'd like to be able to report that this stress on computation ends when students reach college. But, alas, even in calculus, linear algebra, and differential equations courses (courses required by many different majors) there is the same mind-numbing tendency toward routine computational problems. Students are accustomed to this approach; textbooks must appeal to a wide clientele and are therefore bland; and teaching this way places fewer social, intellectual, and time demands on professors. (The latter fact is significant since university mathematics professors receive little, if any, recognition for being effective instructors; they are, however, rewarded if they publish, even when what they publish adds only the tiniest of nuances to the obscure niceties of their narrow specialties.) This computational emphasis is all the more unfortunate given the new software that takes the drudgery out of such processes as evaluating definite integrals and inverting matrices. (I exclude upper-level undergraduate courses and graduate mathematics courses from this indictment, but the generally higher level of instruction there is of little comfort to most.)

It's bad form to continue railing repetitively about repetitious instruction, so I'll summarize. Mathematics is thinking—about numbers and probabilities, about relationships and logic, about graphs and change—but thinking nonetheless. This message isn't reaching even many of our best students who continue to see it as "Ambition, Distraction, Uglification, and Derision."

CORRELATION, INTERVALS, AND TESTING

∞

Children with bigger feet spell better. In areas of the South those counties with higher divorce rates generally have lower death rates. Nations that add fluoride to their water have a higher cancer rate than those that don't. Should we be stretching our children's feet? Are more hedonist articles in *Penthouse* and *Cosmopolitan* on the way? Is fluoridation a plot?

Although studies do exist which establish all of these findings, the above responses to them only make sense if one doesn't appreciate the difference between correlation and causation. (Interestingly, the philosopher David Hume maintained that in principle there is no difference between the two. Despite some superficial similarity, however, the issues he was getting at are quite different from the present ones.) There are various kinds and various measures of statistical correlation, but all of them indicate that two or more quantities are related in some way and to some degree, not necessarily that one causes the other. Often the changes in the two correlated quantities are both the result of a third factor.

The odd results above are easily explained in this way. Children with bigger feet spell better because they're older, their greater age bringing about bigger feet and, not quite so certainly, better spelling. Age is a factor in the next example as well since those couples who are older are less likely to divorce and more likely to die than are those from counties with younger demographic profiles. And those nations that add fluoride to their water are generally wealthier and more health-conscious, and thus a greater percentage of their citizens live long enough to develop cancer, which is, to a large extent, a disease of old age.

For most, the definitions of the actual measures of correlation are less important than the above simple distinction between correlation and causation. But too often people become mesmerized by the technical details of correlation coefficients, regression lines, and curves of best fit and neglect to step back and think about the logic of the situation. The phenomenon reminds me of people (and I'm one of them —see the end of the *game theory* entry) who buy a new computer or word processor to speed their work and then spend an inordinate amount of time obsessing over the details of the software and the writing of "shortcut" programs, each of which requires a few hours to compose but whose invocation saves only three or four keystrokes.

"Everybody seems to be buying them. The whole country is crazy about these things." "They only talked to 1,000 people. So how do they know?" As these contrary quotes might suggest, the idea of a random sample is another simple statistical notion whose importance isn't always fully appreciated. Without such a sample, a swarm of personal testimonials and "everybody says so"s may mean little, and with one, a surprisingly small number of respondents can be conclusive. Based on observations of such a sample, a confidence interval is a numerical strip (which varies slightly from sample to sample) designed so that with a specified probability (usually 95%) it contains the unknown true value of some population characteristic. Thus, if we poll a random sample of 1,000 people and 43% are in favor of the ideas set forth in the Bill of Rights, then the probability is approximately 95% that the percentage of the entire population favoring these ideas is somewhere in the interval between 40% and 46%, 43% ± 3%.

There are a number of technical matters involved in the calculation and interpretation of confidence intervals, but, as in the case of correlation, a knowledge of these isn't required to understand the basic ideas. In fact, becoming engrossed in the niceties of confidence interval estimation may blind one to the limited scope of this method. It's not that 1,000 people aren't sufficient to give us this ±3% confidence interval; they are. It's rather that such estimation is very sensitive to how a problem is framed or how a question is phrased. If, in the example above, the ideas had been *identified* as coming from the Bill of Rights, the responses would likely be quite different. The beliefs, attitudes, and intentions of the respondents make it impossible to carelessly substitute one phrasing of a question for an extensionally equivalent one.

The testing of statistical hypotheses is another notion from statistics that requires no particular technical apparatus to appreciate. You make an assumption, you design and perform an experiment to test it, then you perform some calculations to see if the results of the experiment are probable enough given the assumption. If they are not, you throw out the assumption and sometimes provisionally accept an alternative hypothesis, statistics being more a matter of disconfirming propositions than of confirming them.

There are two sorts of error that may be made in applying this procedure: A Type I error occurs when a true hypothesis is rejected, and a Type II error occurs when a false hypothesis is accepted. This constitutes a distinction which is useful in less quantitative contexts. For example, when the issue is one of governmental money disbursements, the stereotypical liberal tries to avoid Type I errors (the deserving not receiving their share), whereas the stereotypical conservative is more concerned with avoiding Type II errors (the undeserving receiving more than their share). If the punishment of criminals is the topic, the cartoon conservative is more involved with avoiding Type I errors (the deserving or guilty not receiving their due), whereas the liberal frets more about avoiding Type II errors (the undeserving or innocent receiving undue punishment).

The FDA must evaluate the relative probabilities of a Type I error (not okaying a good drug) and a Type II error (okaying a bad drug). Managers concerned with quality control must balance Type I errors (rejecting a sample with very few defective items) and Type II errors (accepting a sample with too many defectives). In these and many other situations, the logic of statistical testing, even when it doesn't result in precise numbers, will stand us in good stead.

Statistics, more than most other areas of mathematics, is just formalized common sense, quantified straight thinking. Skeptical attitudes toward statistics (such as Benjamin Disraeli's "Lies, damn lies, and statistics" or, my favorite, "Truths, half-truths, and statistics") are fully warranted, but shouldn't blind us to the subject's indispensability. Renouncing its use would be making a Type I error (or, if one's feet are small, a Type I eror).

DIFFERENTIAL EQUATIONS

Analysis (calculus and its descendants) has been one of the predominant branches of mathematics since Newton and Leibniz invented it, and differential equations are at its core. The subject has traditionally been the key to understanding the physical sciences, and in the deeper issues it suggests, it has been the source for many of the concepts and theories that constitute higher analysis. It is also one of the most essential practical tools scientists, engineers, economists, and others have for dealing with rates of change. (One of its lesser-known attributes is the minor pleasure it affords some sophomore—sophomoric?—math majors and others when they refer to their "diffy-q" course. I once knew a pharmacy student who took the course because he liked saying "diffy-q" all the time.)

The derivative of a quantity (see the entries on *calculus* and *functions*) is a mathematical function or formula that measures the rate of change of the quantity. If a ball is thrown into the air, for example, the derivative of its height is its velocity. If C is the cost of producing X widgets, the derivative of C with respect to X is the marginal cost of producing the Xth widget. Higher derivatives—e.g., the derivative's derivative—measure how fast the rates of change are themselves changing. Equations that involve derivatives should perhaps be called derivative equations, but are instead called differential equations. And just as the idea in solving an algebraic equation is to determine from conditions on a number what that number might be, so to solve a differential equation is to determine from conditions on the derivative (and higher derivatives) of a changing quantity what that quantity might be at any given time. Simplistically put, the subject of differential equations is concerned with methods and techniques for finding the value of a quantity at any time when what is known is how fast it and related quantities are changing.

Some examples of circumstances leading to simple differential equations: Nicolae heads due west out of Bucharest at noon and maintains a steady pace of 50 miles per hour; determine his location at any given time that afternoon. A large holding tank contains 100 gallons of brine in which are dissolved 200 pounds of salt; if pure water is allowed to run into the tank at the rate of 3 gallons per minute and the mixture, kept uniform by stirring, runs out at 2 gallons per minute, how long will it take to reduce the amount of salt in the tank to 100 pounds? A rabbit runs due east at 4 miles per hour, and a dog, 1,000 yards north of the rabbit's initial position moves directly toward it at 5 miles per hour; find the path the dog takes. An idealized elastic string is bound at both its ends and plucked; find its position at any later time. [The equation describing this last situation is an important one from physics; for those interested, it is $Y''(X) + KY(X) = 0$, where $Y''(X)$ (alternate notation: d^2Y/dX^2) indicates the second derivative of $Y(X)$, the deviation of the string from its rest position at any given point X along its length.]

The distinction between the first and second derivative of a quantity is important in many contexts. If the quantity in question is a distance or a height, its first derivative is its velocity and its second is its acceleration. You may not naturally gravitate to problems in physics, so consider instead the following common news story: The TV anchor reports in his resonant voice that some economic index is still rising, but not as rapidly as it was last month. Probably unbeknownst to him, he is saying that the derivative of the index is positive (the index's rate of change is positive), but the second derivative of the index is negative (the rate of change of the index's rate of change is negative). The rise is "topping out." Going considerably further down this road leads to collections of differential equations relating the values, rates of change, and rates of rates of change of various economic indicators. These collections constitute an econometric model and can be further manipulated to gain an insight into the real world.

In this and other applications of differential equations, we are often interested in more than one variable; sometimes we know not how fast a quantity changes with respect to time, but how fast it changes with respect to something else; many situations can only be described by collections of interrelated differential equations. The mathematical advances developed over the last 300 years to deal with these problems are among the chief glories of Western civilization. Newton's laws of motion, Laplace's heat and wave equations, Maxwell's electromagnetic theory, the Navier-Stokes equations for fluid dynamics, and Volterra's prey-predator systems are just a few of the many fruits of these techniques and ideas (most of whose authors, sadly, are unknown to the average educated person).

In recent years mathematical research has veered away from these classical areas of "diffy-q." The emphasis now is more on numerical approximation and computing and less on traditional methods involving limits and infinite processes.

E

Even more universal than the well-known pornographic novel *The Story of O* or Kafkaesque stories of Everyman K is The Story of E. (Bizarre beginning, I know, but everybody is entitled to a little peculiarity.) Ranking with π in mathematical importance and usually written in unpretentious lowercase (à la e. e. cummings, to complete a ridiculous authorial triangle), e is roughly equal to 2.718281828459045235360287471352662497. What follows is a sketch of a couple of its properties.

The number e was introduced by the Swiss mathematician Leonhard Euler in the middle of the eighteenth century, and its standard definition may be perplexing at first. The number is defined to be the limit of the sequence of terms $(1 + 1/N)^N$ as the integer N grows larger and larger. When N is 2, the expression above is $(1 + 1/2)^2$, equivalently $(3/2)^2$ or 2.25; for N equal to 3, it is $(1 + 1/3)^3$, equivalently $(4/3)^3$ or 2.37; for succeeding values of N it is $(1 + 1/4)^4$, equivalently $(5/4)^4$ or 2.44, then $(6/5)^5$, $(7/6)^6$, . . . $(101/100)^{100}$, and so on. The value of e is the limit of this sequence of numbers. Thus it is very close to $10{,}001/10{,}000)^{10{,}000}$, which equals 2.718145. It is even closer to $(1{,}000{,}001/1{,}000{,}000)^{1{,}000{,}000}$.

Although abstract, the definition holds the key to e's role in banking and compound-interest calculations. (See also the entry on *exponential growth.*) One thousand dollars invested at 12% is worth \$1,000 ×

(1 + .12) at the end of the year. If the money is compounded semi-annually, it is worth \$1,000 × (1 + .12/2) at the end of six months (because 12% interest per year is equivalent to 6% interest per half year), and it is worth [\$1,000 × (1 + .12/2)] × (1 + .12/2), or \$1,000 × $(1 + .12/2)^2$, at the end of the year. If it's compounded quarterly, it's worth \$1,000 × (1 + .12/4) at the end of one quarter (12% interest per year equaling 3% per quarter), \$1,000 × $(1 + .12/4)^2$ at the end of two quarters, and \$1,000 × $(1 + .12/4)^4$ at the end of the year. If we proceed in this way and compound the money N times per year, it will be worth \$1,000 × $(1 + .12/N)^N$ at the end of the year. Note that, save for a .12 instead of a 1, the last factor is identical to that in the definition of e. A little behind-the-scenes mathematical manipulation shows that daily compounding (N = 365) will result in the money's being worth about \$1,000 × $e^{.12}$ at the end of the year and \$1,000 × $e^{.12T}$ at the end of T years. Incidentally, the exponential function $Y = e^T$, in terms of which this and other instances of exponential growth is expressed, is one of the most significant in mathematics.

There are several other definitions of e, all equivalent of course, and all evidence (after some hocus-pocus) of the naturalness of this number. For this reason and for others connected with calculus, e is the base of the *natural* logarithm function. A little expatiation on the ugly subject of logarithms is needed to clarify this statement. The *common* logarithm of a number is simply the power to which 10 must be raised to equal the number in question. The common logarithm of 100 is 2 because $10^2 = 100$ [alternately, log(100) = 2]; the common logarithm of 1,000 is 3 because $10^3 = 1{,}000$; and the common logarithm of 500 is 2.7 because $10^{2.7} = 500$.

In contrast, the natural logarithm of a number is the power to which e must be raised to equal the number. Thus, the natural logarithm of 1,000 is about 6.9 since $e^{6.9} = 1{,}000$ [alternately, ln(1,000) = 6.9]; the natural logarithm of 100 is 4.6 since $e^{4.6} = 100$; and the natural logarithm of 2 is .7 since $e^{.7} = 2$. It can be shown (mathematician talk for "Believe me") that this latter number, the natural logarithm of 2, plays an important role in finance: Dividing the interest rate one receives into .7 yields the number of years it takes your money to double in value. Thus, rates of 10% and 14% (.10 and .14) will result in your money doubling in approximately 7 years (.7/.1 = 7) and 5 years (.7/.14 = 5), respectively. Instead of explaining this, however, or what exactly is so natural about the natural logarithm, let me mention a couple of ways in which the number e arises in other familiar contexts. (Certainly common logarithms, based as they are on the accidental fact that we have 10 fingers, cannot be termed mathematically natural.)

Imagine an academic department that is about to sequentially interview N candidates for a position as assistant professor. At the end of each interview the department must decide whether or not this is the candidate it wishes to hire. Assume that if it passes over someone, that person cannot later be hired, and if it gets to the last candidate, he or she is hired by default. To maximize its chances of hiring the best candidate, the department settles upon the following strategy. It carefully picks some number $K < N$, interviews and rejects the first K candidates, and then continues interviewing until a candidate better than all of his or her predecessors is found. That person is then hired.

This strategy won't always work. Sometimes the best candidate will be among the first K people rejected; at other times the best candidate will come after the one hired. Still, given the constraints, it can be shown that the optimum strategy possible is to choose $(N \times 1/e)$ as the number K, 1/e being approximately .37 or 37%. Thus, if there are 40 candidates and they're interviewed in random order, the optimum strategy is to reject, out of hand, the first 15 of them (37% of 40), and then to accept the next candidate who is better than all of his or her predecessors. The probability of hiring the best candidate using this method is also, strangely enough, 1/e or 37%. No other strategy yields a better probability of success than 37%. Comparable approaches to choosing a potential spouse make use of the same arguments, although the constraints are less natural in this situation.

Another implausible appearance of the number e results when a secretary randomly scrambles 50 different letters and the addressed envelopes into which they are to be stuffed. If he or she haphazardly fills the envelopes with letters, we might conceivably wonder: What is the probability that at least one letter is placed in the correctly addressed envelope? For no easily explainable reason the answer again involves e. The probability of at least one match is $(1 - 1/e)$, or about 63%. Other scenarios leading to the same probability involve two shuffled decks of playing cards turned over in tandem a card at a time or indiscriminately sorted hats and hat checks at a restaurant.

The number e also pops up in situations where we are interested in record-breaking events of one sort or another. To illustrate, imagine a region of the earth that has had the same weather patterns for eons. This region would still see chance variations in its annual rainfall. If we were to begin keeping track of rainfall in the year 1, we would note that record-breaking rainfalls come less frequently as time goes by. Year 1's rainfall would, of course, constitute a record, and maybe year 4's precipitation would exceed that of the first 3 years and become the new record. We might then have to wait until the 17th year, when the rainfall surpasses that of each of the previous 16 years, for the 3rd record-breaking rainfall. If we were to continue recording the annual precipitation for, say, 10,000 years, we would find that there would have been only about 9 record-breaking rainfalls. If we were to keep at it for a million years, we would probably note only about 14 record-breaking rainfalls.

It is no coincidence that the 9th root of 10,000 and the 14th root of 1,000,000 are approximately equal to e. If after N years there have been R record-breaking rainfalls, the Rth root of N will be an approximation to e, and this approximation approaches e more and more closely as N increases without bound. (The number N must be sufficiently—i.e., humongously—large for the approximations to be accurate.)

Irrational (not expressible as the ratio of two whole numbers and hence not having a repetitive decimal expansion) and transcendental (not the root of any algebraic equation), e is nevertheless ubiquitous in mathematical formulas and theorems. It is intimately involved with trigonometric functions, geometrical figures, differential equations, infinite series, and many other areas of mathematical analysis. The unlikely literary trinity at the beginning was an egregious effort to suggest in a nonmathematical way the pervasive importance of e.

MATHEMATICS IN ETHICS

From Plato to Kant to contemporary philosophers such as John Rawls, ethicists have argued for the necessity of impersonal principles of morality. Mathematics is often derided for being an impersonal subject, but properly understood, this characteristic is part of what makes mathematics so useful, even in ethics, where its invocation might seem odd initially. The great seventeenth-century Dutch-Jewish philosopher Spinoza, among others, exemplified this when he wrote his classic *Ethics* "in the geometric style" of Euclid's *Elements.*

First, a depressing example far removed from Spinoza's stoical "theorems" and rationalist principles. According to UNICEF's 1990 report millions of children die each year suffering from nothing more serious than measles, tetanus, respiratory infections, or diarrhea. These illnesses could be prevented by a $1.50 vaccine, $1 in antibiotics, or 10 cents' worth of oral rehydration salts. UNICEF estimates that $2.5 billion would be sufficient to keep most of these children alive and improve the health of countless others. This amount is equivalent to the annual advertising budget of American tobacco companies (whose products, incidentally, kill almost 400,000 Americans annually, more than were killed during World War II in its entirety), or to the Soviets' monthly expenditures on vodka, or, most offensively, to about 2% of the Third World's own armament spending each year. Furthermore, supplying birth control to those women who want it—estimated conservatively at close to half a billion—would cut population growth by 30 percent, making the financial "burden" of the above minor allocation changes still easier to bear. Family planning and the assurance that their children would live would further cut the population rate, population growth being fastest in countries with high infant and child mortality rates. The arithmetic isn't complicated, but it is essential to gain a perspective on the situation.

People often respond with distaste to the attaching of numerical values to lives or to the making explicit of various trade-offs. Balancing cost vs. value in medical care or price vs. impact in environmental protection is always an unpleasant task. There are times, however, when not being quantitative is a kind of false piety which can only make obscure and thus more difficult the choices we must make. It is here that probability theory and operations research can play a significant role. There are also other times, I hasten to append, when the economic arithmetic that is appropriate is more Cantorial (as in biblical and infinite-set-theoretical); i.e., when every life is of infinite worth and thus not less valuable than the sum of any number of other infinitely valuable lives, just as $\aleph_0$, Cantor's first infinite cardinal number, is equal to $\aleph_0 + \aleph_0 + \ldots \aleph_0$. (See the entry on *infinite sets.*)

The necessity for trade-offs isn't always appreciated sufficiently, but neither is it an uncommon concern. Rather than discussing it further, I'd like to point out a couple of nonstandard "applications" of mathematics to the field of ethics. The first is mathematical only in an extended sense of the term, but extending the popular conception of mathematics is one of my intentions.

Assume that a society must make a major policy decision and that a positive decision on the policy carries with it much future risk. If the policy is adopted there will initially be some upheaval—people changing residence, much building and construction, new organizations formed—but the risky policy will lead to an increase in the standard of living for at least 200 to 300 years.

At some indeterminate time after that, however, there will be a major catastrophe, directly attributable to this choice of the risky policy, in which 50 million people die. (Imagine perhaps the decision concerns the disposal of radioactive wastes.) Now, as the English philosopher Derek Parfit has pointed out, it could be maintained that the decision to follow the risky policy was bad for no one. The policy certainly wasn't bad for the people whose standard of living was increased in the centuries before the catastrophe.

Moreover, it wasn't bad for the people who died in the catastrophe since they, those very people who died, wouldn't have been born were it not for the decision to follow the risky policy. This policy, remember, led to some initial upheaval and the consequent altering of when existing couples conceived their babies (and hence of their babies' identities) and also, since different people were brought together, of which pairs became couples and then parents (and hence of their babies' identities). Over the course of centuries these differences multiplied and it could reasonably be assumed that no one alive the day of the catastrophe would have been had not the risky policy decision been taken. The people who die, to reiterate, will owe their lives to the decision.

We thus have an example of a decision, taking the risky course, which it seems is clearly bad—it leads to the death of 50 million people—yet one which is (arguably) bad for no person. What is needed is some impersonal moral principle(s) in whose light we could reject the risky policy.

One candidate for this principle is nineteenth-century philosopher Jeremy Bentham's utilitarian "greatest good for the greatest number." The principle is only schematic, however, and interpreting it precisely so as to create a sort of moral calculus is a skill that thankfully has always eluded moral philosophers. Kant and others have thought that any moral principle ought to be universal, again something useful in the abstract, but not very helpful with details. Ignoring the vast literature connected with these and other approaches to ethical principles, I'll instead introduce here a small bit of quasi-mathematics that is illuminating whatever our (approach to) ethical principles.

Originally phrased in terms of prisoners, the prisoner's dilemma has a generality to it that is difficult to overestimate. Assume that two men suspected of a major crime are apprehended in the course of committing some minor offense. They're separated and interrogated, and each is given the choice of confessing to the major crime and thereby implicating his partner or remaining silent. If they both remain silent, they'll each get one year in prison. If one confesses and the other doesn't, the one who confesses will be rewarded by being let go, while the other one will get a five-year term. If they both confess, they can both expect to spend three years in prison. The cooperative option is to remain silent, while the individualist option is to confess.

The dilemma is that what's best for them as a pair, to remain silent and spend a year in prison, leaves each of them open to the worst possibility, being a patsy and spending five years in prison. As a result, they'll probably both confess and both spend three years in prison.

The appeal of the dilemma has nothing to do, of course, with any interest we might have in the criminal justice system. Rather, it provides the logical skeleton for many situations we face in everyday life. (See the entry on *game theory.*) Whether we're businessmen in a competitive market, spouses in a marriage, or superpowers in an arms race, our choices can often be phrased in terms of the prisoner's dilemma. There isn't always a right answer, but the parties involved will be generally better off as a pair if each resists the temptation to double-cross the other and instead cooperates with or remains loyal to him or her. If both parties pursue their own interests exclusively, the outcome is worse for both of them than if they cooperate. Adam Smith's invisible hand ensuring that individual pursuits bring about group well-being is, at least in these situations, quite arthritic.

The two-party prisoner's dilemma can be extended to circumstances where there are many people, each having the choice whether to make a minuscule contribution to the public good or a larger one to his or her own private gain. This many-part prisoner's dilemma is useful in modeling situations where the economic value of "intangibles" such as clean water, air, and space is an issue. Since to a considerable extent almost all social transactions have an element of the prisoner's dilemma in them, the character of a society is reflected in which transactions lead to cooperation between parties and which don't. If the members of a particular "society" never behave cooperatively, their lives are likely to be, in English philosopher Thomas Hobbes's words, "solitary, poor, nasty, brutish, and short."

Is taking the cooperative option in prisoner's dilemma situations a consequence of any moral theory? Not as far as I know. In fact, a strong case can be made to at least sometimes choose the individualist option in the dilemma; it's not irrational or immoral to defend oneself. Choosing the cooperative option, like many other actions, is neither demanded nor prohibited by any standard ethical theory. And this brings me to my last remark.

Any moral theory, properly formalized and quantified, is subject to the limitations imposed by Gödel's first incompleteness theorem (see the entry on *Gödel*), which states that any sufficiently rich formal system must always contain statements that are neither provable nor disprovable. Given this, we have theoretical support for the common-sense observation that there will always be actions neither demanded nor forbidden by our principles, whatever they are, and even when they are supplemented by our own idiosyncratic fears, values, and commitments. This may be taken as a mathematical argument for the necessity of "situation ethics" and demonstrates the *insufficiency* of an exclusively axiomatic approach to ethics.

EXPONENTIAL GROWTH

∞

The sequence of numbers 2, 4, 8, 16, 32, 64, . . . grows exponentially while the sequence 2, 4, 6, 8, 10, 12, . . . grows linearly. Adapting the ancient story, we note that if 2 cents are placed on the first square of a checkerboard, 4 cents on the second, 8 cents on the third and so on, the last square will have almost $200 million billion on it (2^{64}¢ is approximately $\$1.8 \times 10^{19}$). By contrast, 2 cents on the first square, 4 cents on the second, 6 cents on the third, and so on will result in $1.28 on the last square. In general, a sequence grows exponentially (or geometrically) if its rate of growth is proportional to the amount of the quantity present—i.e., if each number in the sequence comes from multiplying its predecessor by the same factor. A sequence grows linearly (or arithmetically) if its rate of growth is constant—i.e., if each number in the sequence comes from adding the same factor to its predecessor.

Of course, the factor by which you get from one term to the next in a sequence needn't always be two. For example, if the $1,000 you deposit today earns 10% per year, it will grow each year by a factor of 1.1 (or 110%) and be worth $1,100 ($1,000 × 1.1) next year. In two years it will be worth $1,210 ($1,000 × 1.1 × 1.1), and in three years $1,331 ($1,000 × 1.1 × 1.1 × 1.1). Thus the sequence $1,000, $1,100, $1,210, $1,331, . . . is an exponential one and after N years the original deposit's worth in dollars will be $1{,}000 \times 1.1^N$. If the money were to earn 10% simple interest, the original deposit would be worth in successive years \$1,100, \$1,200, \$1,300, and after N years its worth in dollars would be 1,000 + 1,000(.10)N or, equivalently, 1,000 + 100N.

Like money in a compound-interest account, populations (whether of people or bacteria) tend to grow exponentially, and like money earning simple interest, food production tends to increase only linearly. The early-nineteenth-century British economist Thomas Malthus put these two observations together and concluded that poverty and famine were unavoidable. The argument is flawed and can be attacked at a number of points, but the clean way in which it articulates the situation is admirable and illustrative of the difference between exponential and linear growth.

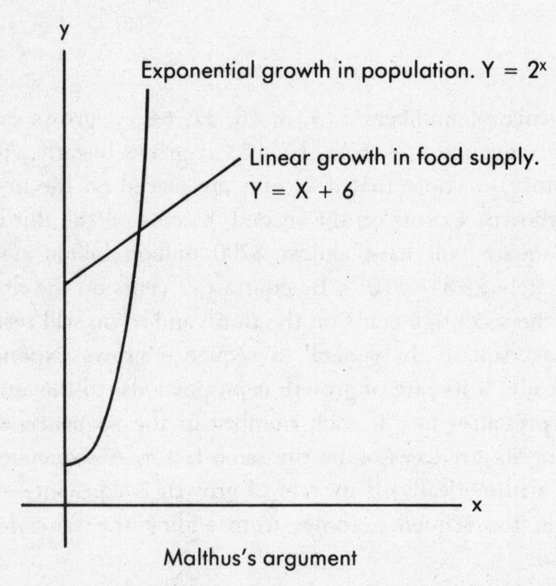

Malthus's argument

Not only does exponential growth easily outstrip the linear variety; it also ultimately outpaces quadratic, cubic, and general polynomial growth. Another look at the two sequences with which I began might help clarify this bit of mathematical jargon. The sequence 2, 4, 8, 16, 32, 64, 128, . . . grows exponentially, its Nth term being equal to 2^N. The sequence 2, 4, 6, 8, 10, 12, 14, . . . displays linear growth, its Nth term equaling 2N. Consider now the sequence 1, 4, 9, 16, 25, 36,

49, ... It demonstrates quadratic growth ("quad" to indicate the terms are squared), and its Nth term is equal to N^2. The Nth term of the cubic sequence 1, 8, 27, 64, 125, 216, . . . is N^3, while the Nth term of the sequence 7, 42, 177, 532, . . . is $2N^4 + 5N$.

The exponential sequence 2^N eventually grows more rapidly than any of these other sequences, more rapidly, in fact, than any polynomial sequence (one whose terms grow as powers of N: N^2, N^3, N^4, N^5, and so on). Note that if $N = 30$, for example, 2^N is 1,073,741,824, while N^4 is only 810,000. Avoiding exponential growth rates is of particular importance when designing computer procedures. (See also the entries on *complexity* and *sorting*.) Those procedures whose required time for solution grows exponentially with the size of the problem posed generally take too long to be of much practical use. If the problem contains a lot of data or is big in some other way, one might have to wait millennia for an answer. Even the speed of supercomputers is no match for such exponential time growth. On the other hand, those procedures whose solution time grows linearly, or at least polynomially, with the size of the problem posed usually yield a solution quickly enough to be of some use.

Let me switch gears and end this brief discussion of exponential growth with a variation in which the sequence shrinks exponentially. The terms of such a sequence are obtained from their predecessors by multiplication by a number less than 1. Thus the sequence 1000, 800, 640, 512, 409.6, . . . manifests what is called exponential decay since each term in it is 80% of its predecessor (the Nth term after the first equaling $1{,}000 \times .8^N$).

An important example of exponential decay is provided by the dissipation of radioactive elements. From a knowledge of the rate of decay of such elements, one can calculate their half-lives (the length of time required for half of the substance to decay). Calculations of this sort are the basis for so-called carbon dating. The idea behind the method exploits the simple fact that all living things contain a known concentration of radioactive carbon in them, and when they die, this radioactive carbon disintegrates. By measuring how much remains, the age of the charcoal or cloth or whatever may be calculated.

Of course, we needn't travel to such arcane realms to encounter exponential decay. The rate at which the taste of Juicy Fruit gum diminishes is also exponential. Another example even closer to home—

my home: It's been said that for every equation one adds to a popular book on mathematics or science, one cuts the potential audience in half. That is, $A = B$ and $X = Y$ will get rid of 75% of those of you who might otherwise have gone beyond this point. The warning could also be phrased, "The number of readers decreases exponentially with the number of equations." I hope it isn't true.

FERMAT'S LAST THEOREM

Many people who doodle with numbers have discovered that the numbers 3, 4, and 5 have the interesting property that $3^2 + 4^2 = 5^2$. Some no doubt have discovered other sets of three numbers with this same property. Two further examples are the sets 5, 12, 13 and 8, 15, 17, which satisfy the equations $5^2 + 12^2 = 13^2$ and $8^2 + 15^2 = 17^2$. It has been proved that there are infinitely many of these so-called Pythagorean triples.

Because the property is such a simple and natural one, mathematicians have wondered whether it might be generalized. In particular, they have investigated whether there are sets of three whole numbers X, Y, and Z such that $X^3 + Y^3 = Z^3$. They have not found any such numbers. They've searched for numbers X, Y, and Z which constitute a solution for $X^4 + Y^4 = Z^4$ and haven't found any. In fact, neither mathematicians nor doodlers have ever found any set of three numbers X, Y, and Z and any number N greater than 2 that together satisfy the equation $X^N + Y^N = Z^N$.

The great seventeenth-century mathematician Pierre Fermat (who should be remembered for his more substantial contributions to number theory, analytic geometry, and probability) wrote that there was a good reason for this failure: There do not exist whole numbers X, Y, and Z such that $X^N + Y^N = Z^N$ for some number N greater than 2. Fermat proved this for $N = 3$, and on a page of a classic Greek text on number theory he noted that he possessed an elegant proof of the general theorem, but that the margin of the page was too narrow for him to reproduce it there. His proof was probably flawed, but it was never found, and no one in the intervening three centuries has succeeded in providing an alternate proof of the general result that has come to be known as Fermat's last theorem.

There have been many partial results (the value of N for which there are no solutions is continually being extended) and in recent years a few proposed proofs have nearly established the result. The primary mathematical importance of the theorem is not intrinsic to the theorem (it is, after all, something of a curiosity), but lies rather in all the algebraic number theory that was discovered and invented in an effort to prove it. The theorem is more like a coal cinder in the eye than a real diamond or, to switch from a carbon-based metaphor to a silicon-based one, more like a piece of sand in an oyster whose irritating presence results in a pearl.

(A more substantive theoretical question was posed at the turn of the century by the German mathematician David Hilbert as part of his famous list of unsolved problems. In addition to queries on Cantor's continuum hypothesis, the consistency of arithmetic, and other abstract matters, Hilbert asked whether there existed a method which could be used in all cases to determine whether or not arbitrary polynomials in several variables—such as $3X^2 + 5Y^3 - 21X^5Y = 12$—had whole-number solutions. Recent advances in logic have shown that there can be no such general method.)

Although it hasn't yet been proven, the consensus is that Fermat's last theorem is true. Still, if it happens to be false, all that is required to demonstrate its falsity and disprove the theorem will be a single counter-example: a set of three numbers X, Y, and Z and some number N greater than 2 such that $X^N + Y^N = Z^N$. Any suggestions for numbers to try?

[One parting piece of logic chopping whose omission might be especially rewarding to the reader: The above paragraph shows that because of its form, if Fermat's last theorem is false, then it is disprovable. So if it is not disprovable, it is true. Thus if it is undecidable (neither provable nor disprovable), it is true. We may conclude that *if* Fermat's last theorem is an undecidable proposition of arithmetic (and there are many—see the entry on *Gödel*), then it is true. This hypothetical reasoning sheds no light on the truth or falsity of the theorem. It is intriguing, however, that from a meta-mathematical knowledge of arithmetic's inadequacy, we could deduce the truth of a simple arithmetic statement.]

MATHEMATICAL FOLKLORE

The phrase "mathematical folklore" sounds a little strange on first hearing, something like "computer fairy tales" or "electronic parables," but mathematics, like other disciplines, has its own yarns and stories which, though often apocryphal, constitute a shared frame of reference and help to define an ethos.

Many of the theorems, examples, and principles discussed herein have short narratives attached to them (Pythagoras' mysticism, Euler's bridges, Fermat's last theorem, Russell's paradox), and in this entry I want only to sketch a few more archetypal stories. "Sketch" and "few" are operative words. I make no claim to being exhaustive (which is sometimes just a euphemism for being exhausting).

Let me start with two time-honored stories about Archimedes, the greatest mathematician, physicist, and inventor of the ancient world. While soaking in a bathtub, it is said, he discovered the eponymously named principle that a body immersed in a fluid is buoyed up by a force equal to the weight of the fluid displaced. Exhilarated by this realization, he ran naked through the streets shouting, "Eureka! Eureka!" (Greek for "I have found it! I have found it!"). The second story about Archimedes also emphasizes his devotion to knowledge. Absorbed in thought over a mathematical diagram sketched in the sand, he was oblivious to the Roman soldier beside him who had ordered him to desist. The soldier thereupon killed him, an act which encapsulates as well as any other the relation between ancient Greek and Roman culture.

Considerably purer in his mathematical tastes than Archimedes, the twentieth-century English mathematician G. H. Hardy took great pride in the worldly uselessness of the theory of numbers. The following interchange between him and his protégé, Indian mathematician Srinivasa Ramanujan, is well known. Visiting Ramanujan in his hospital room, Hardy mentioned that 1,729, the number of the taxi which had brought him, was rather a dull number. To this Ramanujan responded almost immediately, "No, Hardy! No, Hardy! It is a very interesting number. It is the smallest number expressible as the sum of two cubes in two different ways." ($9^3 + 10^3 = 1^3 + 12^3 = 1{,}729$.)

The theme of ineptitude or at least disinterest in mundane affairs also underlies many of the stories about the founder of cybernetics, MIT mathematician Norbert Wiener. Reportedly Wiener's eyesight and/or memory were so bad that a graduate student was assigned the task of making sure he reached his various destinations. Another anecdote about Wiener illustrates the strain of elitism that seems to characterize many mathematicians. Once he was teaching a graduate course and soon discovered, or so a common version goes, that only a student in the front row was following every detail of his presentations. Wiener responded thereafter by speaking directly to this student. One day the student was absent, however, and when Wiener didn't see him in the front row, he left immediately, muttering that nobody was in class.

The intimidating potential of mathematics and mathematicians is illustrated by an exchange between the prolific eighteenth-century Swiss mathematician Leonhard Euler and the French philosopher and encyclopedist Denis Diderot. Before a theological discussion in which he would have fared badly, Euler demanded that Diderot counter an irrefutable yet irrelevant mathematical formula: "Monsieur, $(a + b)^n/n = X$; therefore, God exists. Respond!" Diderot was stupefied and remained silent. (See the entry on *QED*.)

Mathematicians are seldom thought of as romantics. Yet in 1832 the brilliant twenty-one-year-old French algebraist Evariste Galois died in a duel over a prostitute. Alfred Nobel, the inventor of dynamite and founder of the Nobel Prize, was reported (again, the story comes in many versions) to have stipulated that there be no Nobel Prize awarded in mathematics in order to retaliate against his wife's lover, Mittag-Leffler, a likely winner at the time of the prizes' inception.

There are not an abundance of nasty, personal vendettas between mathematicians. What ones there are usually have a significant mathematical component. For example, the bad feelings between the nineteenth-century German mathematician Leopold Kronecker and Georg Cantor, the founder of set theory, revolved in large part around the two men's differing conceptions of the infinite. Kronecker had a finitist Pythagorean view of mathematics and pronounced, "God made the integers and all the rest is the work of man." Cantor, on the other hand, dealt with all sorts of transcendental collections and constructions. Kronecker's assaults on the brilliant but hypersensitive Cantor may have been a factor in the latter's breakdowns and ultimate commitment to a mental institution. Similar, albeit more kindly expressed antagonisms exist between pure and applied mathematicians, between algebraists and analysts, between logicians and all other mathematicians.

More typical is mathematical folklore that involves charismatic figures about whom mathematicians collect anecdotes. Their subsequent retellings are relished like old jokes (and are one of the few good reasons to attend mathematics conferences). Kurt Gödel, for example, is said to have resisted becoming a U.S. citizen for several years because he found a logical contradiction in the Constitution.

Another standard example concerns John von Neumann, who was referred to by some as the smartest person who ever lived. Posed to him was the problem of a bird who flies back and forth between two approaching trains. The bird flies 150 miles per hour and the trains, initially 540 miles apart, travel toward each other at 80 and 40 miles per hour, respectively. The question is: How far does the bird fly before it's crushed between the trains? The plodding way to solve the problem is to calculate the lengths of the bird's successive flights between the collision-bound trains and then to add up the terms of the resulting series. The easy way is to note that the trains meet after 4.5 hours (540 miles/120 miles per hour), and thus the bird's total travels are 4.5 hours × 150 miles per hour = 675 miles. When von Neumann blurted out 675 almost immediately, the poser laughed and remarked that von Neumann knew the trick, to which he is said to have replied, "What trick? What's easier than summing the series?"

Many contemporary stories seem to perpetuate the view that mathematicians are a breed apart, either nerdlike imbeciles (such as the proverbial mathematician who says A, writes B, means C, when the real conclusion is D) or lightning calculators and irrelevant obscurantists. In a classic quote Archimedes maintained that given a fulcrum, a long enough lever, and a place to stand, he could move the earth. The citation suggests the theoretical nature, practical power, and transcendental longings of mathematics and mathematicians. The mathematical idea expressed, the notion of proportion, is a seminal one, as is the notion of recursion, which is the idea expressed in the next story. Contrast, however, the ancient ideal with the more modern stereotype.

A psychologist asked an engineer what he would do if a small fire broke out and there was a pitcher of water on the table. The engineer dutifully replied that he would douse the flames with the water. The psychologist then turned to the mathematician and asked him what he would do if a small fire broke out and there was a pitcher of water on the windowsill. The mathematician replied that he would move the water pitcher from the windowsill to the table and in this way reduce his problem to the previously solved one.

FRACTALS

∞

Imagine you're at the base of a barren mountain. If you were to walk up and down the mountain, you might estimate that the distance you'd walked was approximately 10 miles. Now, what if a 200-foot-tall giant were to take the same path to the summit and back. He might walk only 5 miles. He would be so tall that he would step right over small hillocks without having to go up and down them the way you would. By contrast, imagine an insect crawling up and down the same route. It might walk 15 miles since it would have to go up, over, and down rocks and small boulders that we would merely step over.

Likewise, suppose a tiny amoeba-sized animal were to wriggle its way along the same trail and back. It might travel 20 miles since it would have to go up and down tiny crevices and bumps in rocks and pebbles that even an insect would just step right over. Thus we come to the somewhat odd conclusion that the distance up and down the mountain depends to a large extent on who's doing the traveling. So too does the surface area of the mountain, the amoeba-sized animal finding it a considerably more spacious domain to roam around in than does the giant, who strides right past the smaller minutiae of the surface. The bigger the climber, the shorter the distance. The bigger the climber, the smaller the surface area. This is a characteristic of a fractal, to which the side of a mountain is a good approximation.

A tree's trunk, to cite another standard example of a fractal, branches into a characteristic number of branches which, in turn, each branch into the same number of smaller branches which likewise each break up into the same number of yet smaller branches until we arrive at the twig level. What does this have in common with the surface of a mountain?

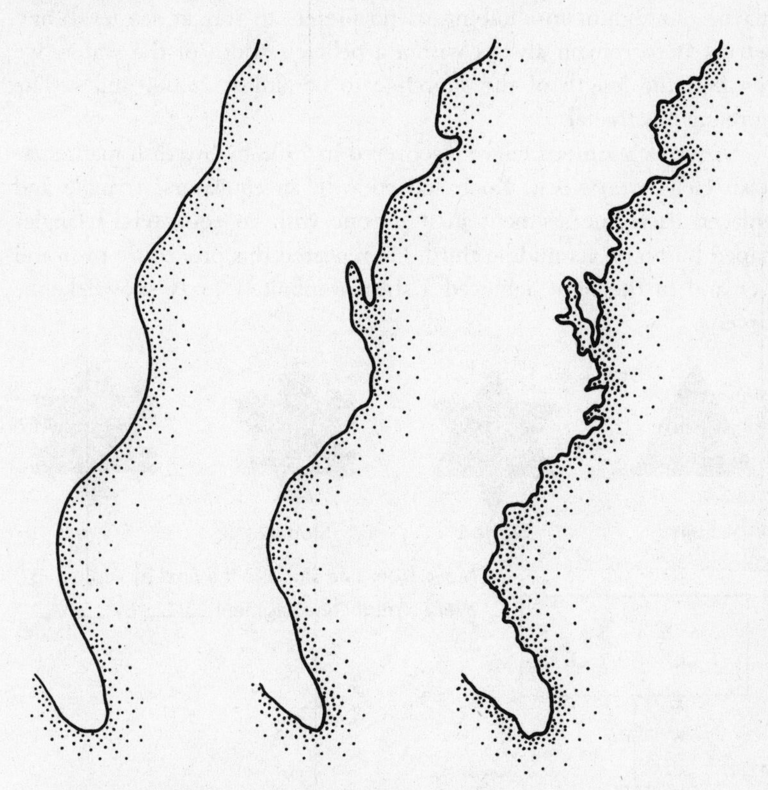

Increasingly fine views of the eastern coast of the United States.

Before we get to a definition, consider a coastal shoreline, yet another example due to mathematician Benoit Mandelbrot, the discoverer of fractal geometry. If we estimate the length of the eastern shoreline of the United States from a satellite, for example, we might come up with a figure of 2,500 miles or so. If, instead, we use detailed maps of the United States, which show the many capes and inlets along the shore, we may increase our estimate of the length of the shoreline to 7,500 miles. If we had nothing to do for a year and decided to walk from Maine to Miami always staying within a yard or two of the Atlantic, the distance we would walk might be closer to 15,000 miles. We would trace not only the capes and inlets on the standard maps but the even smaller juttings and indentations which don't appear on the maps. Finally, if we can convince an insect to walk along the coast (maybe our mountain-climbing friend prefers to stay at sea level) and instruct it to remain always within a pebble's width of the water, we may find the length of the shoreline to be almost 25,000 miles. The shoreline is a fractal.

So too is a famous curve discovered in 1906 by Swedish mathematician Helge von Koch. Koch started with an equilateral triangle and replaced each line segment in it by one with an equilateral-triangle-shaped bump on its middle third. He repeated this procedure over and over and in the limit achieved a strange infinitely fuzzy snowflakelike curve.

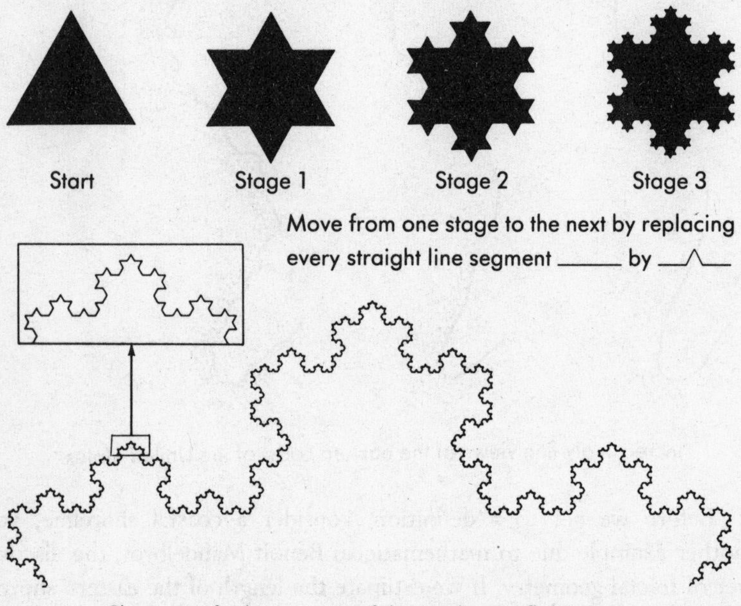

Close-up of a later stage of the Koch snowflake curve

And what is a fractal? It's a curve or surface (or a solid or higher-dimensional object) that contains more but similar complexity the closer one looks. The shoreline, for example, has a typical jagged shape at whatever scale we draw it; i.e., whether we use satellite photos to sketch the whole coast or the considerably more detailed information obtained by a person walking along some small section of it. The surface of the mountain looks roughly the same, whether seen from a height of 200 feet by the giant or close up by the insect. The branching of the tree appears the same to us as it does to birds, or even to worms or fungi in the idealized limiting case of infinite branching. Likewise for the Koch curve.

Blowup of a part of a fractal, due to Benoit Mandelbrot

Moreover, as Mandelbrot has stressed, clouds are not circular or elliptical, tree bark is not smooth, lightning does not travel in a straight line, and snowflakes are most certainly not hexagons (neither do they resemble Koch curves). Rather, these and many other shapes in nature are near fractals and have characteristic zigzags, push-pulls, bump-dents at almost every size scale, greater magnification yielding similar but ever more complicated convolutions. There is even a natural way to assign a fractional dimension to these shapes, the fractals used to model coastlines having dimensions between 1 and 2 (more than a straight line but less than a plane), while those used to model mountain surfaces having dimensions between 2 and 3 (more than a plane, but less than a solid). NASA photos indicate that the fractal dimension of the earth's surface is 2.1, compared with 2.4 for that of Mars's "woollier," more convoluted topography. Coined by Mandelbrot in 1975, the term "fractal" is an apt expression for *frag*mented, *fract*ured self-similar shapes of *fract*ional dimension.

Besides being a boon to computer graphics, where they are used to depict realistic-looking landscapes and natural forms, fractal-like structures are turning up frequently whenever fine structure is analyzed—on the surfaces of battery electrodes, in the spongy interior of intestines and lung tissue, in the variation of commodities prices over time, or in the diffusion of a liquid through semi-porous clays. With their beautiful and intricate complexity at all levels and scales of magnification, fractals are playing an increasingly important role in chaos theory (see the entry on *chaos*), where they can be used to describe a system's collection of possible trajectories. Their grotesque elegance is also apparent in purely mathematical contexts. A plane, for example, is partitioned into regions according to whether one or another root of an equation will eventually be obtained via a standard Newtonian method. The borders between these sections are staggeringly complex fractals.

Novelists too may someday find that fractal analogues in "psychic space" are helpful in capturing the fractured yet nevertheless coherent structure of human consciousness, whose focus can shift instantaneously from the moment's trivia to timeless verities and then back again, somehow preserving the same persona at the various levels. (See the entry on *human consciousness, its fractal nature.*) In this regard, the verbatim transcripts of ordinary conversations are quite revealing. The stops, starts, ellipses, bizarre syntax, vague references, unmotivated digressions, and sudden changes of direction are nothing like the sanitized "linear" version which usually emerges in print. There may be ways in which the above notions could be useful in cognitive psychology as well. The difficulty of a field of study, for example, might be looked upon as a fractal with brighter and/or more knowledgeable people taking larger cognitive steps over the tiny difficulties that others must patiently climb up and over.

FUNCTIONS

∞

The notion of a function is a very important one in mathematics since it captures in a formal way the idea of a correspondence between one quantity and another. The world is full of things which depend on, are a function of, or are associated with other things (in fact, a case could be made that the world simply *consists* of such relations), and we're confronted with the problem of establishing a useful notation for this mathematical dependence. The following examples illustrate one common notation. Other ways to indicate these linkages involve graphs and tables. (See the entry on *analytic geometry*.)

Consider a small workshop which produces chairs. Its costs are \$800 (for equipment, say) and \$30 per chair produced. Thus the relationship between the total cost, T, and the number of chairs produced, X, is given by the formula $T = 30X + 800$. If we want to stress the dependence of T on X, we say that T is a function of X and symbolically denote this association by $T = f(X)$. If 10 chairs are produced, the cost is \$1,100; if 22 are produced, the cost rises to \$1,460. The function f is the rule which associates 1,100 with 10 and 1,460 with 22, and we indicate this by writing $f(10) = 1{,}100$ and $f(22) = 1{,}460$. What is $f(37)$?

The Celsius temperature C can be obtained from the Fahrenheit temperature F by subtracting 32 from the latter and multiplying the difference by 5/9. In equational form we have $C = 5/9(F - 32)$. Thus a chilly 41 degrees Fahrenheit translates into an equally chilly 5 degrees Celsius, while a balmy 86 degrees Fahrenheit becomes an equally balmy 30 degrees Celsius. By substituting the Fahrenheit temperature into the formula in this way, we can always find the corresponding Celsius temperature. As before, if we wish to stress the dependence of C on F, we say that C is a function of F and denote this relationship by C = h(F). (The graphs of this function and the previous one are straight lines.) The function h is the rule which associates 5 with 41 and 30 with 86, and this correspondence is denoted symbolically by writing h(41) = 5 and h(86) = 30. What is h(59)?

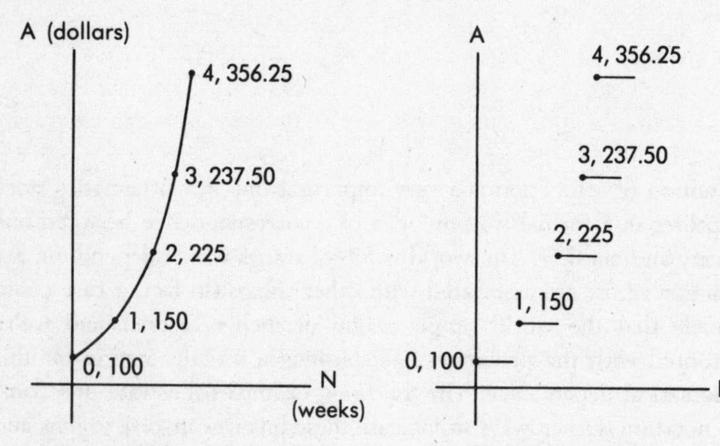

Graph of amount, A, owed as an exponential function of the length of time owed, N
$A = 100(1.5)^N$

Graph of amount owed if fees remain constant between weekly increases

Or imagine that you're a loan shark. You lend $100 to someone and inform him that the amount he owes you will rise 50 percent each week. Checking with your accounting associates, you understand that the amount, A, that your friend owes you after N weeks is equal to $100 \times (1.5)^N$; i.e., $A = 100(1.5)^N$. Clearly A is a function of N, a fact we indicate by A = g(N) (or by the function's graph, an exponentially rising curve). It's clear that g(1) = 150, g(2) = 225, and g(3) = 337.50. (If you're kind and only increase your loan fees at weekly intervals, the graph will consist of a sequence of exponentially rising steps.)

Unless you habitually think of physics as the plural form of physic, consider the following. You throw a ball up into the air with an initial velocity of 80 feet per second from a rooftop which is 200 feet high. Take Newton's word for it, and assume that the height H of the ball above the ground is given by the formula $H = -16T^2 + 80T + 200$, where T is the number of seconds that elapse from the instant you release the ball. Since the height depends on the time, H is a function of T and we write $H = s(T)$. Substituting $T = 0$ into the formula confirms that initially $H = 200$. Two seconds later $T = 2$, and we find by substituting 2 into the same formula that $H = 296$. Thus $s(0) = 200$ and $s(2) = 296$. What is s(5) and why is it less than s(2)?

The functions h, g, and s above are, respectively, linear, exponential, and quadratic functions, while $p(X) = 3\tan(2X)$ and $r(X) = 7X^5 - 4X^3 + 2X^2 + 11$ are termed trigonometric and polynomial functions. But functions needn't always be defined by formulas or equations and needn't always involve numbers. For example, if m(Helen) = red, m(Rebecca) = yellow, m(Myrtle) = brown, m(George) = black, m(Goldilocks) = gold, and m(Peter) is undefined, it's not too difficult to guess that the function m is the rule which associates each person with his or her hair color and that Peter is bald. Thus m(X) merely denotes X's hair color. Likewise, p(X) might be defined to be the author of X, and q(X) might be taken to be the state capital nearest to X. If so, p(*The Cherry Orchard*) = Chekhov, and q(Philadelphia) = Trenton, N.J.

In the examples given, the number of chairs produced, the Fahrenheit temperature, the number of weeks until the debt is repaid, the number of seconds after release of the ball, and the name of a person are termed independent variables. The total cost, the Celsius temperature, the amount of money returned, the ball's height, and the person's hair color are termed dependent variables. Once the value of an independent variable is fixed, it uniquely determines the value of the dependent variable, and the latter variable is said to be a function of the former.

Variants of the notation are used when we have quantities which depend on more than one other quantity—i.e., when we have functions of two or more variables. If $Z = X^2 + Y^2$, for example, then $Z = 13$ when $X = 2$ and $Y = 3$; if we wish to stress the dependence of Z on X and Y, we write $Z = f(X,Y)$ and $13 = f(2,3)$.

The notation for functional dependence is bookkeeping, but essential bookkeeping. It enables us to express relationships in capsule form.

Without it much of the flexibility and power of mathematical analysis can't be easily tapped.

[Answers to questions: f(37) = 1,910; h(59) = 15; s(5) = 200 and s(2) = 296; the ball is on its way up at T = 2, on its way down at T = 5.]

GAME THEORY

∞

Many if not all life situations can be viewed as games if "game" is interpreted broadly enough. (Of course, if we interpret "broadly enough" broadly enough, many life situations can be viewed as zucchini too, but this exceeds almost everyone's linguistic tolerance.) It's not surprising, then, that the mathematical discipline of game theory plays an essential role in the way economic, military, and policy planners frame their choices and decide on their strategies. Invented by John von Neumann about fifty years ago with these applications in mind, it can also clarify more personal decisions and trade-offs.

Game theory is most useful when there is an element of bluff involved and when probabilistic strategies are therefore required. In games with perfect information such as checkers or chess, there is always an optimal deterministic strategy, and moves needn't be random or secret. Although much is known about games of this sort, the existence of a winning strategy for them doesn't necessarily mean it can be found in "real time." To this day the optimal strategies for chess and checkers are unknown, although ones for simpler games such as tic-tac-toe are familiar to kindergartners.

A game situation arises when two or more players are each free to select from a set of possible options or strategies. These choices in turn result in various outcomes—payoffs or penalties of different magnitudes. Each player has preferences among these outcomes. Game theory is concerned with the determining of players' strategies, costs and benefits, and equilibrium outcomes.

Rather than developing the principles of the subject, however, let me describe a typical game situation which lends itself to probabilistic strategies. Consider a pitcher and batter facing each other. The pitcher can throw a curveball, a fastball, or a screwball. Prepared for a fastball, the batter averages .300 against a curveball (i.e., gets a hit 30% of the time), .400 against a fastball, and .200 against a screwball. If he's expecting a curveball, however, he averages .400 against such pitches, .200 against fastballs, and .000 against screwballs. And if he's prepared for a screwball, his averages against curveballs, fastballs, and screwballs are, respectively, .000, .300, and .400.

		Batter prepares for a		
		Curveball	Fastball	Screwball
Pitcher throws	Curveball	.400	.300	.000
	Fastball	.200	.400	.300
	Screwball	.000	.200	.400

Probabilities of a hit

On the basis of these probabilities, the pitcher must decide which pitch to throw and the batter must anticipate this and prepare accordingly. If the batter decides to prepare for a fastball, he can certainly avoid a .000 batting average. If he does this repeatedly, however, the pitcher will throw nothing but screwballs and hold him to a .200 batting average. The batter might then decide to prepare for screwballs, which, if the pitcher continued to throw them, would give him a batting average of .400. The pitcher might anticipate this and throw curveballs, which, if the batter continued to prepare for screwballs, would result in a .000 average. It's clear that their reasoning could cycle endlessly about in this manner.

Each player needs to devise a general probabilistic strategy. The pitcher must decide what percentage of his pitches should be curveballs, fastballs, and screwballs and then throw them *randomly* according to these percentages. The batter must likewise decide what percentage of the time he must prepare for each type of pitch and then do so *randomly* with these percentages in mind. The techniques and theorems of game theory enable us to find the optimum strategies for each player in this game and for a wide variety of other games. It turns out that the solution to this particular idealized game is for the pitcher to throw screwballs 60% of the time and curveballs the remaining 40% of the time, and for the batter to prepare for fastballs 80% of the time and for screwballs the other 20%. If they follow these optimum strategies, the hitter's batting average will be .240.

The familiar game of chicken does not admit of such a clean resolution. A typical version involves two teenagers who drive their cars toward each other at a high rate of speed. The first one to swerve loses face; the other is the victor. If they both swerve, it's a wash. If neither does, they crash. More quantitatively, teenagers A and B each have the choice of swerving or not. If A swerves and B doesn't, the result is, let's assume for illustration's sake, 20 points for A and 40 points for B. The scoring is reversed if B swerves and A doesn't. If they both swerve, the payoff is 30 points for each, while if neither swerves, the "payoff" is 10 points apiece. Like the prisoner's dilemma (see the entry on *mathematics in ethics*), the situation is quite general and is not limited to teenage cretins. Like the prisoner's dilemma also is the fact that individuals seeking only to maximize their personal payoff don't.

It doesn't take much imagination to see that there are many situations in business (labor conflicts and market battles), sports (virtually all competitive contests), and the military (war games) that can be modeled in a game-theoretic way. One relatively new example is provided by the devices that reveal the phone number of the person calling you, the caller's option of preventing her number from being revealed to you, and your choice in each case of whether or not to answer the call. Although most applications commonly invoke disturbing words like "battle," "war," and "contest," this vocabulary isn't essential. The subject could just as easily be called negotiation theory as game theory. Its principles are applicable in so-called non-zero-sum games where one player's payoff is not necessarily balanced by the other's negative outcome, in intimate negotiations (battle of the sexes?), in more "communal" games such as maintaining a just government, and even in games where one of the players is Nature or the environment.

A useful tool, the technical apparatus of game theory shouldn't be allowed to obscure the assumptions which go into any particular negotiation or contest. ("We had to destroy the village in order to save it.") It's depressingly easy to get caught up in constructing payoff matrices and in calculating the expected consequences of various strategies and not to think through the enabling suppositions and one's ultimate objectives. I suffer from the anesthetic effects of such technophilia, and this book was written in part as an expiation.

GÖDEL AND HIS THEOREM

The mathematical logician Kurt Gödel was one of the preeminent intellectual giants of the twentieth century, and, assuming the survival of the species, will probably be one of the few contemporary figures remembered in 1,000 years. A number of recent books about him notwithstanding, this judgment is not a result of hype or an incipient fad (although it is made infinitesimally more acceptable by the similarity among the words "God," "Gödel," and "Godot"). Neither is it, despite a tendency for all disciplines to foster professional myopia, a case of mathematicians' self-congratulation. It's simply true.

Who was Kurt Gödel? The biographical outline is uncomplicated. Born in 1906 in Brünn (in what is now Czechoslovakia), he went to the University of Vienna in 1924 and remained there until he emigrated to the United States in 1939. He lived in Princeton, New Jersey, and worked at the Institute for Advanced Study from that time until his death. During the 1930s and early 1940s, he discovered results in mathematical logic that revolutionized understanding of that subject. The research also shed light on related areas of mathematics, computer science, and philosophy.

His most famous achievement, the so-called first incompleteness theorem, shows that any formal system of mathematics that includes a modicum of arithmetic is incomplete: There will always be true statements that will be neither provable nor disprovable within the system no matter how elaborate it is. No one will ever be able to write out a list of axioms and then rightfully claim that all of mathematics follows from these axioms (even if the axioms fill up a whole tablet, or an entire library of a million books, or quadrillions of silicon chips exhausting all the sand in the Sahara). By distinguishing rigorously between statements within a formal system and meta-statements about the system, by utilizing clever recursive definitions, and by assigning numerical codes to the statements of arithmetic, Gödel was able to construct an arithmetical statement which "says" of itself that it is unprovable and thus establish his result. (Conceivably Boris Pasternak had Gödel's theorem in mind when he wrote, "What is laid down, ordered, factual, is never enough to embrace the whole truth.")

An alternate proof of the theorem due to American computer scientist Gregory Chaitin employs notions from complexity theory (see the entry on *complexity*). In this approach the undecidable arithmetical proposition "says" via numerical code that a certain random sequence of bits is of a complexity greater than that of the given formal system. This fact is known to be true from meta-level considerations, but in order for the proposition to be provable within the system, the system would have to generate a sequence of bits that had greater complexity than it itself had. By the definition of complexity this is impossible.

Germane is the so-called Berry paradox, which directs: "Find the smallest whole number which requires in order to be specified more words than there are in this sentence." Examples such as the number of hairs on my head, the number of different states of the Rubik cube, and the speed of light in millimeters per century, each specify, using fewer than the number of words in the given sentence, some particular whole number. The paradoxical nature of the Berry task becomes apparent when we note that the Berry sentence specifies a particular whole number which, by definition, it contains too few words to specify. Although both proofs of Gödel's theorem play off of well-known paradoxes, the liar paradox for the standard proof and Berry's paradox for Chaitin's, the incompleteness theory itself is not in the least a paradox. It's strange, but it is real and unproblematic mathematics.

It should also be mentioned that, despite strenuous efforts to show the contrary, the theorem does not point to any fundamental cleavage between brains and machines. Both are subject to limitations and constraints that, in principle at least, are quite similar. Even the ability to "stand back" may be given to a machine by formalizing its meta-language and, if necessary, its meta-meta-language.

Gödel's other fundamental results deal with, among other topics, intuitionism, provability and consistency in mathematics, recursive functions, and, later in his life, cosmology. Using many of the ideas and constructions of the first theorem, his second incompleteness theorem states that no reasonable system of mathematics can demonstrate its own consistency. We can only assume the consistency of such a system; we can't prove it without making assumptions even stronger than that of consistency.

Gödel lived an ascetic personal life, the only outward evidence of emotion or personality being his long marriage and his periodic bouts with depression, for which he was hospitalized on a number of occasions. Slightly less solitary as a young man, he was peripherally involved with the Vienna circle of philosophers, although he was unsympathetic to their positivism. Except for this and his later friendship with Einstein at the Institute, however, there was little social interaction between him and his contemporaries. His contacts with mathematicians were limited primarily to papers, correspondence, and telephone conversations. There was in Gödel none of the passionate involvement of Bertrand Russell or the robust humor of Einstein.

He did have other intellectual interests, however. His work led him to the conviction that numbers existed in some domain independent of man, and that the explanation of mind was not mechanistic since mind was separate from and irreducible to matter. Coming from a Lutheran background, Gödel wasn't conventionally religious, but always maintained his theism and the possibility of a rational theology. He even attempted to construct a variant of the medieval ontological argument that God's existence was somehow a consequence of our being able to conceptualize Him. A truly great logician, he must have known intense intellectual exhilaration. Still, crude sensualist that I am, I couldn't help wishing he had been a little happier in a more visceral sense—better health, a child, a love affair, something physical.

Gödel died in Princeton on January 14, 1978, of (according to the death certificate) "malnutrition and inanition" brought about by "personality disturbances."

GOLDEN RECTANGLE, FIBONACCI SEQUENCES

∞

If an ancient Greek mathematician were transported through time to a modern office supply store, one of the many things that might impress him would be the 3 by 5 index cards. After admiring the familiar straightedges and compasses for a while and then the leather briefcases and fancy calculators, he might wander back to the cards again, the 5 by 8 variety having caught his eye this time. The reason for his fascination with these cards (I know it's quite possible he couldn't care less about them, but let's pretend) would be that their dimensions approximate those of the golden rectangle, a shape our Greek mathematician and his contemporaries considered most appealing.

Before returning to these cards, let me first define the closely related notion of a golden section, a ratio that many have found to be quite harmonious. Imagine that we have before us a line segment AB and we wish to divide it at some interior point C. The point C we choose might be the midpoint of AB, but this is a boring choice, so let's assume that C will divide AB into a longer part AC and a shorter part CB. The Pythagoreans would counsel us to choose C so that the ratio of the whole line to the longer part is equal to the ratio of the longer part to the shorter part—i.e., AB/AC = AC/CB. If C is chosen in this way, C is said to divide AB in the golden section, and this golden ratio (of the whole to the longer or, equivalently, of the longer to the shorter) can be computed to be about 1.61803 to 1 (except perhaps on Wall Street, where the golden ratio is more likely to be price/earnings).

C is said to divide AB in the golden ratio if AB/AC = AC/CB.

Any rectangle whose length-to-width ratio equals the golden ratio is said to be a golden rectangle.

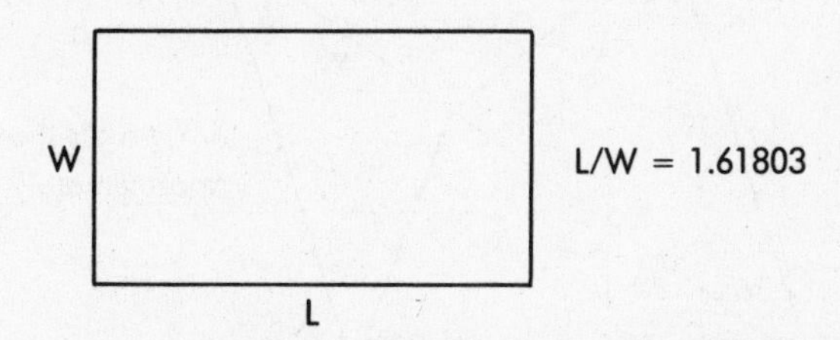

Index cards measuring 3 by 5 come close since 5/3 = 1.6.

A golden rectangle is defined to be any rectangle whose length-to-width ratio is the same as this golden ratio. Not surprisingly, the Parthenon in Athens can be framed by a golden rectangle, as can many of the smaller areas within it. Much other Greek art made use of the proportions of the golden rectangle, as have subsequent works from da Vinci to Mondrian and Le Corbusier. The famous Fibonacci sequence of numbers 1, 1, 2, 3, 5, 8, 13, 21, 34, 55, 89, 144, 233, . . . bears an unexpected relation to these golden rectangles and provides the link with the index cards mentioned above. The sequence is defined by the fact that each term in it (except for the first two) is the sum of its two predecessors: 2 = 1 + 1; 3 = 2 + 1; 5 = 3 + 2; 8 = 5 + 3, 13 = 8 + 5. (See also the entry on *recursion.*)

Remembering that the golden ratio is approximately 1.61803 (the number has a nonrepeating infinite decimal expansion) and performing a little long division, we notice (what can be proved with a little algebra) that the ratio of a term in the Fibonacci sequence to its predecessor approaches this ratio. For the case of 3 by 5 cards, the ratio 5/3 = 1.66666; for 5 by 8 cards, the ratio is 8/5 = 1.6; 13/8 = 1.625; 21/13 = 1.615384; 34/21 = 1.61905; and so on. If the Greeks were right, there might be a bigger market for 8-by-13-inch tablets than there is for the 8½-by-11-inch size.

The Fibonacci sequence also manifests itself outside of office supply stores. In the sunflower, for example, the number of spirals to the left and the number to the right are generally adjacent numbers in the sequence. Likewise, the number of rabbits in succeeding generations seems to follow a Fibonacci pattern, while the shell of the chambered nautilus can be generated Fibonaccially (to coin an awful adverb).

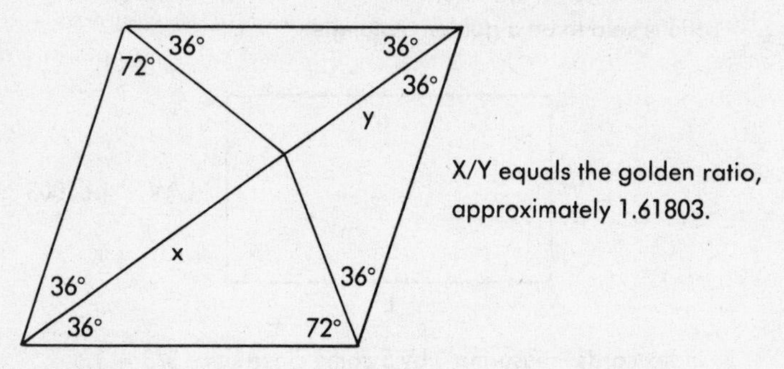

Penrose's two simple shapes (a dart and a kite) and how they fit together

Part of a nonperiodic covering of the plane with Penrose's two simple shapes

The golden rectangle and the static harmony it exemplifies is typical of classic Greek geometry, while the Fibonacci sequence, which dates from about 1200, suggests the slow awakening of a more quantitative and numerical approach to mathematics. Both bespeak a placidity that seems a little incongruous with our more fractured, spiky era whose appropriate mathematical emblem is chaos theory.

But mathematics is no respecter of pompous historical pronouncements, and in the early 1970s the English mathematical physicist Roger Penrose discovered a new instance of the golden ratio with a more modern flavor. He found two simple shapes (one shaped like a kite, the other like a dart), copies of which are capable of covering the plane in a nonperiodic manner and whose relative dimensions are described by the golden ratio. Furthermore, and this is the modern part, the shapes cannot cover the plane periodically.

GROUPS AND ABSTRACT ALGEBRA

∞

Abstract algebra and modern geometry (see the entry on *non-Euclidean geometry)* were products of the nineteenth century, and both subjects helped to change our view of the nature of mathematics. Mathematics is no longer thought to be solely concerned with eternal verities, but is recognized as often being merely a way of deriving the consequences of various sets of axioms. As Bertrand Russell put it: "Pure mathematics consists entirely of such asseverations as that, if such and such a proposition is true of anything, then such and such another proposition is true of that thing." One of the most important "anythings" in abstract algebra is a group.

Not a collection of gregarious mathematicians, a mathematical group is an important kind of abstract algebraic structure. Because groups are abstract, I'll present some examples of them before I give their definition. Consider first the set of whole numbers, positive, negative, and zero, and the operation of addition. Observe what may not, at first glance, seem especially significant: The sum of any two numbers is a number; the equation $(3 + 9) + 11 = 3 + (9 + 11)$ holds true and, more generally, how we associate additions does not affect the sum; there is a number, 0, such that $X + 0 = X$ for any number X; for any number X, there is another number which, when added to X, gives 0: $6 + (-6) = 0$, $(-118) + 118 = 0$, and so on.

Or consider the twelve objects *0, 1, 2, 3, . . . 10,* and *11.* The operation on these objects is like addition on numbers except that when a sum exceeds 11, it is replaced by its remainder upon division by 12 (the operation is usually called addition modulo 12). Thus $8 + 7 = 3$, and $(6 + 5) + 9 = 8$. It's easy to check that the first three properties mentioned above hold for this set of objects and this operation as well. The fourth property also holds but isn't quite so obvious. What must be added to 7 to give *0*? There is no -7 here, but there is a *5*, and $7 + 5 = 0$. One can verify that every object has an "inverse" which, when added to it, yields *0*.

Sticking with numbers, ponder the four objects *1*, *2*, *3*, and *4*, but this time let the operation be like multiplication except that when a product exceeds 4, it is replaced by its remainder upon division by 5 (multiplication modulo 5). Thus, $4 \times 3 = 2$. Again the natural analogues of the four properties hold. The product of two objects is an object; the equation $(3 \times 2) \times 4 = 3 \times (2 \times 4)$ holds true and, more generally, how we associate products is immaterial; there is an object, *1* in this case, such that $1 \times X = X$ for any object X; for any object X, there is another object which, when multiplied by X, gives *1*. For *2* it is *3* since $2 \times 3 = 1$, while for *4* it is *4* itself since $4 \times 4 = 1$.

The above three sets with their respective operations are all groups, but I'll defer the general definition a bit longer to demonstrate that groups needn't have anything to do with numbers or numberlike objects. The elements of this next group are permutations (or rearrangements) of three objects which I'll imaginatively call A, B, and C. Continuing with the colorful appellations, I'll label the first permutation P_1. It doesn't permute A, B, and C at all, but simply leaves them alone. P_1 is called the "identity" permutation and plays a role analogous to 0, *0*, and *1* above, the identity elements of their respective groups. The next permutation, P_2, switches A and B, but leaves C unmoved. P_3 switches B and C, leaving A unmoved, while P_4 switches A and C and leaves B unmoved. P_5 rearranges A, B, and C as C, A, and B, while P_6 permutes A, B, and C to read B, C, and A. Furthermore, these P's affect any arrangement to which they are applied, so that the effect, for example, of P_4 on B, A, C is the arrangement C, A, B, the elements in the first and third positions being switched.

To check that the analogues of the four properties hold, we need an operation on this set of permutations P_1 through P_6. Let the operation be "sequential application": Apply one permutation to the *results* of another to yield a third "product" permutation. What, for example, would the product P_2*P_3 be? To find out, let's trace what happens to A under this product. Since it switches A and B, P_2 puts A in B's position. P_3 then switches the B and C positions, placing A in C's position. Tracing B next, we note that P_2 places it in A's position, where P_3, switching the B and C positions, leaves it. As for C, P_2 leaves it alone, after which P_3 moves it into B's position. Thus P_2*P_3 rearranges A, B, and C as B, C, and A, which is the effect of P_6 alone. Hence $P_2*P_3 = P_6$.

You might want to verify other products in this mathematical shell game. (Then again you might not.) For example, $P_2*P_2 = P_1$, or $P_4*P_6 = P_2$. The point is not to inundate you with bookkeeping (there are notational devices to help with this), but to demonstrate that these six permutations under the sequential application operation, *, satisfy the four properties above: The product of any two permutations is another permutation; how we write these permutations does not affect their product, as $P_i*(P_j*P_k) = (P_i*P_j)*P_k$; there is a permutation, P_1, such that $P_1*P_i = P_i$ for any permutation P_i; and for any permutation P_i, there is another permutation, P_j, such that $P_i*P_j = P_1$.

Not unanticipated at this point, the formal definition of a group is: any set with an operation defined on it which satisfies the above four properties. There are innumerable other examples of groups, many of them geometric. The above group of permutations may, for example, be interpreted as a collection of reflections and rotations of a triangle, each permutation of the vertices corresponding to a movement of the triangle. (Groups like this which are essentially the same—corresponding identity elements, corresponding operations—but differ only in the names of their elements and operations are termed isomorphic.) Groups also arise in knot theory and are helpful in the analysis and classification of knots and braids. They play a role in many other areas of mathematics, in crystallography, and in the study of quarks and quantum mechanics. In most cases the group elements are some sort of action—a permutation, a flexing, a function of some sort. Even the various twistings of the faces of a Rubik cube constitute a group.

What saves this all from being merely a kind of mathematical botany (definition and classification, but with little depth) is that many powerful theorems have been proved about groups, subgroups, quotient groups, and their relation to other abstract structures. To cite one example: If G is any finite group and H any subgroup of G, then the number of elements in H divides the number in G. The power of isolating these abstract structures and proving theorems about them derives in large part from a simple fact. Once a theorem is proved about an abstract group, it applies to any and all of the disparate examples of groups. By focusing on the abstract structure rather than the specific instantiations, we're able to see the mathematical forest rather than its individual trees.

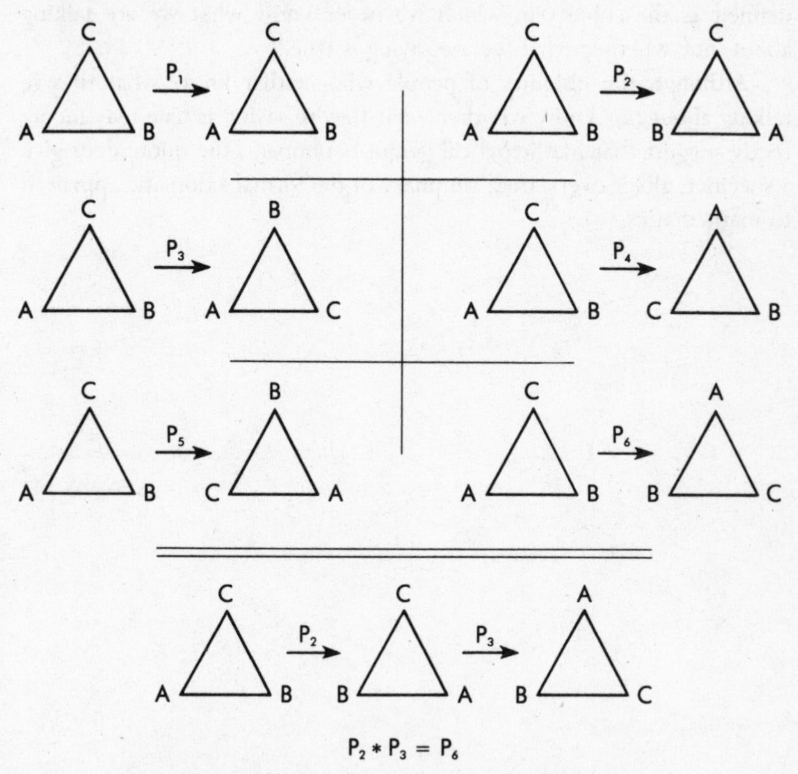

The permutations of A, B, and C can be interpreted as reflections and rotations of a triangle and the * operation as the performance of these movements in succession.

And this, finally, is what abstract algebra is all about. Just as elementary algebra uses variables to symbolize numbers and study their properties, the theories of groups, rings, fields, vector spaces, and other algebraic structures carry the abstraction considerably further. These theories use symbols for sets, operations on sets, structures, functions between structures, and so on, to aid in the proving of general theorems and propositions.

The rest of the Russell quote is an apt close. "It is essential not to discuss whether the proposition is really true, and not to mention what the anything is of which it is supposed to be true. . . . If our hypothesis is about anything and not about some one or more particular things, then our deductions constitute mathematics. Thus mathematics may be defined as the subject in which we never know what we are talking about, nor whether what we are saying is true."

Although the ubiquity of people who neither know what they're talking about nor know whether what they're saying is true may incorrectly suggest that mathematical genius is rampant, the quote does give a succinct, albeit overstated, summary of the formal axiomatic approach to mathematics.

HUMAN CONSCIOUSNESS, ITS FRACTAL NATURE

Fractals are curves, surfaces, or higher-dimensional geometrical figures which have the property of maintaining their characteristic structure under magnification, more of the same sort of complexity being revealed the closer one looks at them. (See the entry on *fractals.*) This self-similarity is suggestive of human consciousness, which seems to have a specific feel to it whether one is thinking logically with some purpose in mind or aimlessly musing or, something having caught one's interest, homing in for a more detailed look, which, in turn, may lead to a more searching examination with yet finer detail or to a return to one's original line of thought.

Serious discussions of international arms reduction and barbershop banter both have a characteristic human "shape" to them, and, as mathematician Rudy Rucker has suggested, the forward movement, horizontal digression, branching, and backtracking at various levels and scales define this human shape and constitute a fractal in a many-dimensional logical space. Since the previous sentence is vague, but reminiscent of the Argentine writer Jorge Luis Borges, I'll try to clarify it with a Borges-like little fiction (whose eponymous protagonist will be called Rucker). The story that follows takes the form of a review of an imaginary book, something similar to which I wish already existed. Reviewing this book which hasn't been written is considerably easier than writing it, however.

RUCKER: A LIFE FRACTAL, by Eli Halberstam
Published by Belford Books, Boston, 3,213 pp., $39.95
Disk Version by Peaches N' Cream Software, Atlanta, $79.95
Reviewed by Paul John Allen

Renowned mathematician Eli Halberstam (recipient of the coveted Fields Medal in mathematics and author of the best-sellers *Mind Matters* and *Chaos, Choice, and Chance*) has written a gargantuan first novel based on Benoit Mandelbrot's esoteric notion of a fractal, and it's safe to say that nothing like it has ever been attempted—even by such as James Joyce or Marcel Proust. It doesn't matter where you begin the book since one doesn't read it so much as wander around in it. There is no conventional story line; rather, there are indefinitely many excursions, all unified by the consciousness of one Marvin Rucker, Halberstam's alter ego.

The 3,213-page tome begins with middle-aged mathematics professor Rucker in his study puzzling through some tedious theorems associated with the well-known NP = P problem. The real novelty, however, is explained in the introduction where the reader (browser) is told that after reading a passage, one can proceed forward linearly, backtrack to a previous passage, or move horizontally by focusing on any major word or phrase in the passage, and then be directed to a further elaboration of it. Sounds simple enough, but the proof is in the doing, or as Rucker rather prosaically thinks to himself on a number of occasions, "God is in the details."

For example, Rucker idly picks his nose while thinking about his theorems, and if the reader chooses to follow up on this, he is directed to a page (on the disk version the alternatives are listed on a menu which appears at the bottom of the monitor) where Rucker's keen interest in proboscis probing is discussed at length. What percentage of people pick their noses? Why do so few people do it in public; yet, in the false privacy of their automobiles why do so many indulge? If you push even further in this direction, there is the memory from a few weeks previous when Rucker, stopped at a red light, saw the elegantly coiffed Mrs. Samaras seated in the BMW across from him, her index finger seemingly deep into her frontal cortex.

Should you tire of this, you can back up and return to Rucker's study, where his young son has just entered, gob-stoppers candy dribbling down his chin. Rucker's about to gently scold him for gumming up his new calculator when he remembers how he himself used to love chewy Chuckles as a youth. Again the reader can proceed with the story or follow up on children's candy, preoccupied parents, or the tone of voice one uses to intimidate kids, each alternative leading to a number of others. The virtue of this arboreal proliferation of digressions is the fragile, evanescent, lifelike feel it lends the book.

Halberstam advises the reader to read only the developments, asides, and vignettes that intrigue him, no more than one fourth of the book at most. The computer version has a little quiz at the end, the answers to which are dependent on which portions the reader has selected. Independently, a few friends and colleagues of mine read the book on a video screen, and, Rashomon-like, our answers to the quiz questions did differ substantially and in the way the computer, which recorded our reading selections, had predicted.

Even in a book as mammoth as this, one can't develop every conceivable fork the various tales might take. It is Halberstam's artistry which overcomes this combinatorial explosion of possibilities and seamlessly binds and weaves the material, creating the illusion of unbounded bifurcation. There are several major stories, among them one dealing with Rucker's complicated home life, another involving a barely legal con game, and a third illustrating in a most intriguing way the salient ideas of complexity theory, the hottest new area in computer science and mathematical logic.

Often at crucial junctures there are few, if any, alternatives. The effect, like a rushing stream, is to suggest the protagonist's single-mindedness at these times. For example, Rucker, an intellectual buffoon vaguely reminiscent of a Saul Bellow character, has dialed a 976 pornography number out of a combination of prurience and curiosity. After getting into the spirit of things, he hears the beep indicating that another call is waiting. He depresses the receiver twice, only to discover his wife calling from the supermarket. Flustered, he tries to end their conversation quickly by telling her that he's on the line with a colleague of his from school, whereupon she exclaims that she must speak with him herself about his wife's choice of a band for their son's upcoming bar mitzvah, and would he hurry up and press the pound key so that the call can be converted to a conference call. He can't, of course, because the lubricious lady on the other line would put a new dent in his already banged-up conjugal relationship.

Despite such narrative twists, it is the almost sentient matrix of diversion, digression, and horizontal movement within the work which vivifies Rucker and his exploits and which most impresses the reader. Details, both big and small, on matters both critical and trivial, tumble forth from this baroque, multidimensional chronicle. To those of us in mathematics, Halberstam seems to be saying that human consciousness —like endlessly jagged coastlines, or creased and varicose mountain surfaces, or the whorls and eddies of turbulent water, or a host of other "fractured" phenomena—can best be modeled using the geometrical notion of a fractal. The definition isn't important here, but unlimited branching and complexity are characteristic of the notion, as is a peculiar property of self-similarity, whereby a fractal entity (in this case, the book) has the same look or feel no matter on what scale one views it (just the main events or finer details as well).

Rather than explain this further (Halberstam doesn't), I'll content myself with observing that by manifesting the inexhaustibility of human rambling and maundering, the book also demonstrates the unity and personal integrity of human consciousness. The structure of the work is virtuosic, and though John Updike needn't worry, the writing is quite serviceable—about all one can wish for given the length. The book carries its didactic burden easily, and for all its sprawl, one comes away from it with a vivid and precise grasp of a surrogate person—Marvin Rucker. The stories are dense with life; they are, in fact, almost incompressible, and further plot summary would be misleadingly reductionistic. The closest literary cousins are Joyce's *Ulysses* and Laurence Sterne's *Tristram Shandy,* but both of these lack the cerebral muscularity of *Rucker: A Life Fractal.*

The book deserves a wide audience, which, unfortunately, it may not attract because of the fear which anything even faintly mathematical engenders in so many. It may be dismissed as a mere technical feat or as mere science fiction or as mere something else, about which our largely innumerate literati know little and toward which they are therefore quite dismissive. The fact that this review is allotted only 1,250 words is some evidence for this possibly paranoid position.

Like Bellow's Herzog, Rucker writes to a motley collection of people, some famous, others infamous, some living, others dead. One of his many "correspondents" is Alexander Herzen, a nineteenth-century Russian writer and liberal dissident. Herzen's famous remark, "Art and the summer lightning of human happiness—these are our only true

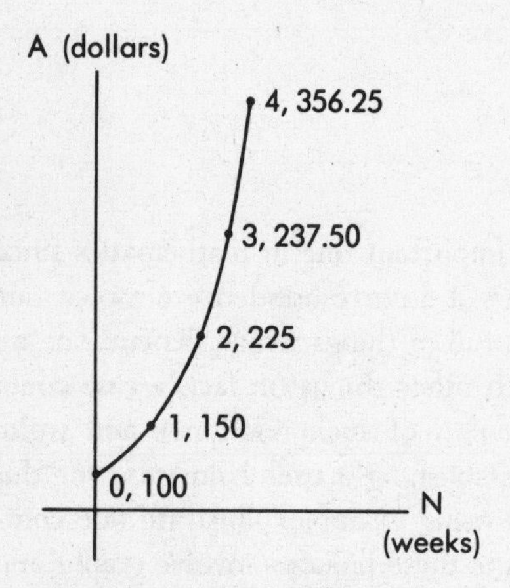

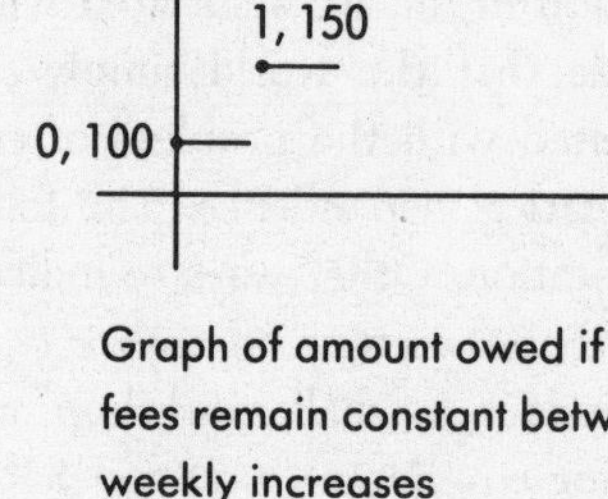

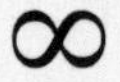

goods," is twice cited by Rucker. Perhaps Halberstam resonated with the juxtaposition of lightning, which has a fractal structure, and art, which in this particular instance does too. In any case, *Rucker: A Life Fractal* delivers the true goods.

HUMOR AND MATHEMATICS

Most people whose school exposure to language was limited exclusively to the study of punctuation would probably not have much of an appreciation for literature. Students of art whose training consisted primarily of drawing rectangular solids might well be oblivious to emotion or beauty in art. Likewise, people whose mathematical education has been restricted to mastering the techniques of computation or to routine courses taught by rote will find talk of humor in mathematics a little strange.

I first became interested in the connection between mathematics and humor when I noticed that mathematicians frequently had a characteristic sense of humor. Possibly because of their training, they tended to interpret statements literally, and these literal interpretations were often incongruous with the standard ones. Since incongruity of one sort or another is a necessary condition for humor, their literalness was often funny as well. "Keep litter in its place," for example, means to leave the stuff on the ground; otherwise it loses its status as litter. If you put it in the garbage can, it's no longer litter, but garbage. The "Lower your voice" sign in the library does not mean that you should converse under the table or that you should speak in a deep baritone. The beauty pageant host who gushes, "Each girl is prettier than the next," is literally saying that they're getting uglier. "I'd give my right arm to be ambidextrous" is funny only if taken literally.

These slightly puerile jokes are typical of mathematicians (more accurately of some mathematicians, including myself). Also typical are a preoccupation with wordplay, self-reference (especially among my fellow logicians), nonstandard models of situations, reversals, reductio ad absurdum, iteration, and a number of other logical and quasi-mathematical notions. Both mathematics and humor are forms of intellectual play (see the end of the entry on *non-Euclidean geometry*), mathematics being on the intellectual side and humor on the playful side of a continuum which stretches between them and passes through puzzles, paradoxes, and brain teasers. Lewis Carroll's injunction to letter writers to lick the envelope and not the stamp is marginally funny and marginally mathematical (inversion of a logical relation) and is thus a representative inhabitant of this middle area.

Furthermore, both mathematics and humor are combinatorial, involving as they do the taking apart and putting together (juxtaposing, generalizing, interpolating, reversing) of ideas just for the fun of it. In humor this is a commonplace, but this same aspect of mathematics is not as well-known, perhaps because it's first necessary to have some mathematical ideas before one can play around with them. Seeing, for example, that a mathematical notion like the knotting of braids has a surprisingly cute connection to another idea in a disparate area, say the symmetries of a geometric figure, requires a certain amount of mathematical sophistication.

Ingenuity and cleverness are hallmarks of both humor and mathematics, as is a spartan economy of expression. Long-windedness is as antithetical to pure mathematics as it usually is to good humor. At the risk of being long-winded myself, I'll note that the beauty of a mathematical proof often depends on its elegance and brevity. (See the proofs of the infinitude of primes, the irrationality of the square root of 2, or the uncountability of the real numbers.) A clumsy proof introduces extraneous considerations and is circuitous or redundant. Similarly, a joke loses its humor if it is awkwardly told, is explained in excessive detail, or depends on strained analogies. Jokes can survive questionable taste, but they must be punchy: AIXELSYD.

Patterns, rules, structures, and logic are essential to both mathematics and humor, although in not quite the same way. In humor, the logic is often perverse or self-referential. For example, there is the case of the voter who when asked by a pollster what were the reasons for the ignorance and apathy of the American public, responded, "I don't know and I don't care." The patterns and structures that are utilized are usually distorted or confused, as they are by the man who noted that olive oil comes from the squeezing of olives and coconut oil from the squeezing of coconuts and then wondered at the source of baby oil. And the relevant rules are misunderstood, as in the case of the two clergymen discussing the sad state of sexual morality. "I didn't sleep with my wife before we were married," one of them declared self-righteously. "Did you?" "I'm not sure," said the other. "What was her maiden name?"

These mistakes and non sequiturs are not random, however, and must make sense on some level (e.g., solar-powered flashlights, or "This sentence is a !!! premature punctuator"). Understanding the correct logic, pattern, rule, or structure is essential to understanding what is incongruous in a given story, to "getting the joke." Conversely, being aware of the elegance, snap, and force of a mathematical proof is essential to a real appreciation of what mathematics is. Again, the uses to which these understandings and appreciations are put in mathematics and in humor are quite different. For example, mathematicians utilize one of their favorite gambits, the logical technique of reductio ad absurdum, to prove propositions. In order to prove S, it is enough to assume the denial of S and from this denial to derive a contradiction or absurdity. Comedy writers use the technique too when they start with an odd premise ("What would happen if . . .") and then develop the joke or story for the sake of the consequent absurdities.

Although the emotional chords differ, there is a posturing and arrogance to research mathematicians which is not dissimilar from that of stand-up comics. The smile elicited by an unexpected twist in a beautiful proof is a rarefied version of the laugh evoked by a good joke's punch line. I close with an easy query: What is a question that contains the word cantaloupe for no apparent reason?

[Note: AIXELSYD is DYSLEXIA spelled backward.]

IMAGINARY AND NEGATIVE NUMBERS

An abbreviated history of number systems can be sketched by considering the solutions to various kinds of algebraic equations. (See also the entry on *Arabic numerals.*) The equation $2X + 5 = 17$ has the positive whole-number solution of 6. No problem here, although we need to go beyond these most user-friendly of numbers to find a root for $3X + 11 = 5$. The solution to the latter equation is a negative number, -2, but taking the plunge and calling -2 a number took a while. The exact status of negative numbers was unclear to mathematicians for a long time, and even today their properties are a bit mysterious to many beginning algebra students. Nobody has much difficulty with negative numbers per se. Fifteen degrees below zero is quite comprehensible both viscerally and intellectually. But why is a negative number times a negative number a positive number?

The answer is a formal one. The product of two negative numbers is defined to be positive so that these numbers might obey the same arithmetical laws as do the positive whole numbers. Money situations illustrate this nicely. Assume you receive \$100 per week from a small annuity and you promptly place it under your mattress. Then in 7 weeks there will be \$700 more under your mattress than there is today ($7 \times \$100 = \$700$), while 5 weeks ago there was \$500 less under your mattress than there is today ($-5 \times \$100 = -\$500$). Assume now, years later, that your annuity has long since ended, and you have been paying out \$100 per week from your cache under the mattress. Then in 8 weeks there will be \$800 less under the mattress than there is today ($8 \times -\$100 = -\$800$), while 3 weeks ago there was \$300 more under your mattress than there is today ($-3 \times -\$100 = \$300$).

So we add the negative whole numbers to our number system. But the integers, positive and negative combined, still aren't sufficient to supply us with a root for $5X - 1 = 7$, whose solution, 8/5, is a fraction (rational number). Again we're hospitable and add all fractions to our number system. Even the rational numbers aren't enough to satisfy our lust for solutions, however. The equation $X^2 - 2 = 0$, for example, has the square root of 2 as a solution, and this number is not rational. Neither is the solution to $4X^3 - 7X + 11 = 0$. Perhaps if we add all these algebraic numbers and all the irrational numbers to our number system, we'll be able to solve every algebraic equation.

Wrong. The simple equation $X^2 + 1 = 0$ has no solution. There is no real number X (rational or irrational) such that $X^2 = -1$ because the square of any real number is greater than or equal to 0. What to do? We invent a new symbol, i, and simply define it (as did Euler, d'Alembert, and others) to be $\sqrt{-1}$, the square root of -1. Thus $i^2 = -1$, and we have a solution to our equation. The letter i originally was meant to suggest the imaginary nature of this number, but with the greater abstraction of mathematics, it came to be realized that it was no more imaginary than many other mathematical constructs. True, it is not suitable for measuring quantities, but it obeys the same laws of arithmetic as do the real numbers, and, surprisingly enough, it makes the statement of various physical laws very natural.

The set of numbers of the form $a + bi$, where a and b are real numbers, constitutes the complex numbers, a number system which includes the real numbers as a subsystem. (The real numbers are simply those complex numbers in which $b = 0$. Thus the real numbers 7.15 and π may be written as $7.15 + 0i$ and $\pi + 0i$, respectively. The number i may be written as $0 + 1i$.) The sum of two complex numbers, say $3 + 5i$ and $6 - 2i$, is defined to be $9 + 3i$. Subtraction is defined analogously, while multiplication and division make use of the fact that $i^2 = -1$. Given this expanded number system, we can prove the fundamental theorem of algebra which states that not only does $X^2 + 1 = 0$ have solutions in the complex numbers, but so, finally, does any algebraic equation. (See the entry on *the quadratic and other formulas.*) The equations $2X^7 - 5X^4 + 19X^2 - 11 = 0$, $X^{17} - 12X^5 + 8X^3 = 0$, and $3X^8 - 26X - 119 = 0$ all have solutions in the complex-number systems. Furthermore, quadratic equations (those in which the variable is squared) have two roots, cubic equations have three, quartic equations have four, and, in general, polynomial equations of degree N (in which the variable is raised to the Nth power) have N roots.

Although the early proponents of imaginary numbers proceeded formally and with only a slight understanding of what they were doing, others soon extended the definition of trigonometric and exponential functions to the domain of complex numbers and generalized mathematical analysis (calculus, differential equations, and related fields) to accommodate the extension. In particular, sense was given to the raising of a number to an imaginary power, and a result is one of the most remarkable formulas in mathematics: $e^{\pi i} = -1$, where e is the base of the natural logarithm function. Writing the formula as $e^{\pi i} + 1 = 0$ relates the five most significant constants in mathematics in a single equation. (I know. So does the equation $e^{0\pi i} = 1$, but since the 0th power of anything is 1, the use of e, π, and i in this instance is vacuous.) These technical advances, including the geometrical interpretation of various operations on complex numbers, paved the way for their indispensable use in electrical theory and other physical sciences. Their development also stimulated the growth of abstract algebra, and in particular the subjects of vector analysis and quaternions.

The number i is evidence that much real progress can result from the positing of imaginary entities. Theologians who have built elaborate systems on much flimsier analogies should perhaps take heart.

IMPOSSIBILITIES—
THREE OLD, THREE NEW

∞

Students often worry too much about finding solutions to problems and seldom appreciate that some of the most interesting problems in mathematics (and elsewhere) don't have any solutions. Three classical construction problems from antiquity as well as three revolutionary results from the twentieth century fit into the latter category.

The constructions require that one use only an unmarked straightedge and a compass to (1) duplicate a cube, (2) trisect an angle, and (3) square a circle. Since these expressions aren't completely self-explanatory (duplicating a cube sounds a little like sleight of hand), let me expand on them. Given a certain line segment, one duplicates a cube by constructing another line segment whose length is such that a cube having this latter segment as a side would have twice the volume of a cube having the original segment as a side. Just doubling the length of the original segment won't work since this will result in a cube having a volume eight times that of the original, not twice it. What it means to trisect an angle is fairly clear: One must divide the angle into three equal parts. The method, however, must work for any angle and not just for certain selected ones. And to square a circle means to construct a line segment whose length is such that a square having this segment as a side would have an area equal to that of the given circle. Remember, too, we want for each problem a method that is exact in principle, not just an approximation.

Not until the nineteenth century when they conclusively demonstrated that these constructions were impossible did mathematicians give up their attempts to perform them. The equations associated with these problems involve either cube roots or the number pi (which satisfies no algebraic equation—see the entry on *pi*), but the numbers and lengths constructible with straightedge and compass were proved to be limited to those definable by square roots (and square roots of square roots). Nevertheless, the mathematics that grew out of these futile attempts is considerably more valuable, both practically and theoretically, than the constructions themselves. Work on analytic curves, cubic and quartic equations, Galois theory, and transcendental numbers all stem, at least in part, from these impossibilities.

This sand-in-the-oyster phenomenon is quite general in mathematics. Recognition of the irrationality of the square root of 2 destroyed the Pythagorean belief that all things were explainable in terms of whole numbers and their ratios but stimulated Eudoxus, Archimedes, and others to develop a theory of irrational numbers. Difficulty deriving Euclid's parallel postulate from the other postulates eventuated in the nineteenth century's discovery (actually it was the discovery of János Bolyai, Nikolai Lobachevski, and Karl Friedrich Gauss) of non-Euclidean geometry. Georg Cantor's set-theoretical oddities near the turn of the century energized the mathematical community and led in part to the foundational work of Bertrand Russell, Alfred North Whitehead, and others.

Tellingly, three of the most significant discoveries of the twentieth century take the form of impossibility statements. Kurt Gödel's incompleteness theorem states that in any axiomatic mathematical system which includes arithmetic there will always be statements that are neither provable nor disprovable within the system; i.e., it is impossible to prove within the system all the truths about the system. This dashed the hopes of those who thought that all the truths of mathematics might be derived from a single axiomatic system. (See the entry on *Gödel*.)

In physics, Werner Heisenberg's uncertainty principle states that it is impossible to determine exactly the position and momentum of a particle at any given time and that the product of the uncertainties in these two quantities always exceeds a certain constant. In addition to its revolutionary impact on physics, the uncertainty principle made adherence to a strictly deterministic philosophy of science much more problematic. Unfortunately, it has also inspired many "parapsychologists." The quotes indicate my quaint conviction that practitioners of a discipline ought to have a discipline to practice. Impressed by some vague formal similarities between quantum mechanics and human behavior—predictive uncertainty, extreme sensitivity to observation, failure of certain logical laws, incommensurability or complementarity of "outlooks"—these parapsychologists have given a scientific gloss to belief in telepathy, psychokinesis, and precognition without providing them with any scientific substance.

Finally, economist Kenneth Arrow's theorem on "social choice functions" states that there is never a foolproof way to derive group preferences from individual preferences that can be guaranteed to satisfy certain reasonable and minimal conditions. In other words, it is impossible to design a voting system that will not in some situations evince serious shortcomings. As with Gödel and Heisenberg, an intellectual ideal had to be abandoned, this time the hope for a universal method of making social choices. (See the entry on *voting systems.*)

Recognizing theoretical impossibility is often a measure of intellectual sophistication. Primitive people and societies can usually solve all their problems.

MATHEMATICAL INDUCTION

∞

Imagine a ladder with infinitely many rungs reaching up into heaven, Plato's penthouse (above his cave), whatever. We can reach this exalted position if we are able to get on the first rung (or some rung) of the ladder, and whenever we're on any given rung (say K), we are always able to climb to the next rung $(K + 1)$. The formalization of this idea is the axiom of mathematical induction, one of the most powerful tools in the mathematician's arsenal. (See also the entry on *recursion.*) An alternate metaphor: an infinite row of dominoes. If one falls, the rest will, but if the first one isn't pushed or if one domino is missing, then they won't all fall.

Consider as an illustration the proof that $1 + 2 + 3 + 4 + \ldots + N = [N(N + 1)]/2$. First pick a number out of the air, say 7, and see if the formula is true for it. Does $1 + 2 + 3 + 4 + 5 + 6 + 7 = (7 \times 8)/2$? The answer is yes, since both sides equal 28. You might also check that the formula holds for 10. Although these and other confirming instances suggest that the formula may be true for all integers, they don't prove it. For this task, we'll use mathematical induction. Does the formula hold true for $N = 1$, or, in terms of the ladder analogy, can we get onto the first rung? Plugging 1 into the formula, we obtain $1 = (1 \times 2)/2$, which is, of course, true. The formula works for $N = 2$ as well since $1 + 2 = (2 \times 3)/2$.

So we're on the ladder. But can we always move from one rung to the next? Assume we're swaying precariously on the Kth rung (for some specific yet arbitrarily chosen K) so that the formula holds for this K: $1 + 2 + 3 + 4 + \ldots + K = [K(K + 1)]/2$. To demonstrate that we can clamber up to the next rung, we must prove that the formula holds for $(K + 1)$—i.e., that $1 + 2 + 3 + 4 + \ldots + K + (K + 1) = [(K + 1)(K + 2)]/2$. We may substitute for the sum of the first K terms, $1 + 2 + 3 + 4 \ldots + K$, in this second equation that which we're assuming to be equal to this sum from the first equation, namely $[K(K + 1)]/2$. Performing this substitution, we're left with $[K(K + 1)]/2 + (K + 1) = [(K + 1)(K + 2)]/2$.

This equation is what needs to be proved, and to do so is now merely a matter of expanding the two sides algebraically and checking that they're equal. (Those for whom the word "merely" is a joke should try to follow the logic of the above, which is considerably more interesting and sophisticated than the algebra anyway.) Having fulfilled both conditions of the induction axiom, we reach mathematical heaven and conclude that the formula is true for all integers N.

Merely establishing instances of a general proposition may sometimes make it plausible, but certainly does not constitute an inductive proof. One can check, for example, that $(N^2 + N + 41)$ is a prime number for many values of N. (Prime numbers cannot be broken into factors as 35 can; $35 = 5 \times 7$.) For $N = 1$, the expression $(N^2 + N + 41)$ is equal to 43; for $N = 2$, it equals 47; for $N = 3$, 53; for $N = 4$, 61; for $N = 5$, 71; for $N = 6$, 83; for $N = 7$, 97; for $N = 8$, 113; and for $N = 9$, 131. In fact every value of N up through 39 yields a prime number for $(N^2 + N + 41)$. The proposition fails for $N = 40$, however.

Mathematical induction may be used to prove any proposition which involves an arbitrary integer N. For example, we might use it to demonstrate that the sum of the angles of a convex polygon of N sides (triangle, rectangle, pentagon, hexagon, and so on) is always $(N - 2) \times 180$ degrees. How we come upon such propositions is a question distinct from how we prove them and doesn't lend itself to a formal analysis. In this regard I note that mathematical induction should be distinguished from scientific induction, which is, at least roughly, the inferring of general empirical laws from specific instances. Theorems proved by mathematical induction are deductive and certain, while conclusions depending upon scientific induction are only probable at best.

For extra credit, you may examine the statement that $(4^N - 1)$ is divisible by 3 for all values of N. It's true for N = 2, for example, since $(4^2 - 1)$ (which equals 15) is divisible by 3. Can you prove the general proposition by induction?

For very much more extra credit, demonstrate that induction really does lead to heaven. Show that there are only two requirements for immortality: Be born, and ensure that whatever the day, you live until the day after.

[For interested parties, the proof of the above proposition: Certainly $(4^1 - 1)$ is divisible by 3. Assume now that $(4^K - 1)$ is divisible by 3. We must show that $[4^{(K+1)} - 1]$ is also divisible by 3. We first note that $[4^{(K+1)} - 1]$ equals $[(4 \times 4^K) - 1]$. Then we use a little algebraic trick and rewrite the latter expression as the sum: $[4 \times (4^K - 1)] + (4 - 1)$, in effect both subtracting and adding 4. Since, by our assumption, $(4^K - 1)$ is divisible by 3, so is $[4 \times (4^K - 1)]$. Finally, since $(4 - 1)$ is divisible by 3, and since the sum of two expressions which are each divisible by 3 is itself so divisible, we conclude that $[4^{(K+1)} - 1]$ is divisible by 3.]

INFINITE SETS

∞

You arrive at the hotel, hot, sweaty, and impatient. Your mood is not improved when the clerk tells you that they have no record of your reservation and that the hotel is full. "There is nothing I can do, I'm afraid," he intones officiously. If you're in an argumentative frame of mind, you might in an equally officious tone inform the clerk that the problem is not that the hotel is full, but rather that it is both full and finite. Explain that if the hotel were full but infinite, there would be something he could do. He could tell the party in room 1 to move into room 2; the latter party he could move into room 3, whose previous occupants would have already been moved to room 4, and so on. In general, the guests in room N would be moved into room (N + 1) for all numbers N. This action would deprive no party of a room yet would vacate room 1, into which you could now move.

Infinite sets have many counter-intuitive properties not shared by finite ones. An infinite set, by definition, may always be put in one-to-one correspondence with a subset of itself; i.e., it has subsets whose members may be paired up one-to-one with the members of the whole set. The hotel scenario above illustrates that if we remove the number 1 from the set of whole numbers, there are just as many numbers remaining (2, 3, 4, 5, . . .) as there are whole numbers altogether (1, 2, 3, 4, . . .). Likewise, there are just as many even numbers as there are whole numbers, and just as many whole numbers that are multiples of

17 as there are whole numbers. The following pairing makes this latter fact clear: 1, 17; 2, 34; 3, 51; 4, 68; 5, 85; 6, 102; and so on.

Some of the oddities associated with infinite sets have been known since Galileo, but the German mathematician Georg Cantor studied them systematically, and largely through his efforts set theory has become the common language of abstract mathematics. Not astonishingly, the topic of infinite sets is a very big one, and thus I will discuss here only a useful distinction due to Cantor—that between countably and uncountably infinite sets. This cleavage plays an important role in mathematical analysis, and the proofs associated with these notions are particularly beautiful.

A set is countably infinite if there is some way to associate or match up its elements (with no leftovers) in a one-to-one fashion with the positive whole numbers. An uncountably infinite set is one whose elements can't be so matched up with the positive integers. The infinite sets we've encountered so far have all been countably infinite, but before I provide an example of an infinite set which is uncountably so, I will sketch a proof of Cantor's which demonstrates that the set of all rational numbers is also countably infinite despite its density and seeming plenitude. That is, there are just as many integers as there are fractions. (See the entry on *rational and irrational numbers.*)

How can one match up the rational numbers (fractions) with the positive integers? We can't just link 1/1 with 1, 2/1 with 2, 3/1 with 3, and so on, because we would be leaving out most rational numbers. There are similar problems with other, more sophisticated attempts. One trick that works is to first consider those rational numbers the sum of whose numerators and denominators is 2. There is only one of these —1/1—and we associate the integer 1 with it. Now consider those rationals the sum of whose numerators and denominators is 3. There are two—1/2 and 2/1—and we associate with them the integers 2 and 3, respectively. After this we examine those rationals whose numerators and denominators sum to 4—1/3, 2/2, and 3/1—and we associate the next available integer, 4, with 1/3, then 5 with 3/1, and ignore 2/2 (and all fractions not in lowest terms). At the next stage, fractions whose numerators and denominators add up to 5, we associate 6 with 1/4, 7 with 2/3, 8 with 3/2, and 9 with 4/1.

We continue in this way, at each stage considering those rational numbers the sum of whose numerators and denominators is N, ordering them as to size, and then associating subsequent integers with them. Eventually every rational number is linked with an integer, and we conclude that the set of fractions is countably infinite. (You might want to check that the integer 13 is associated with the fraction 2/5.)

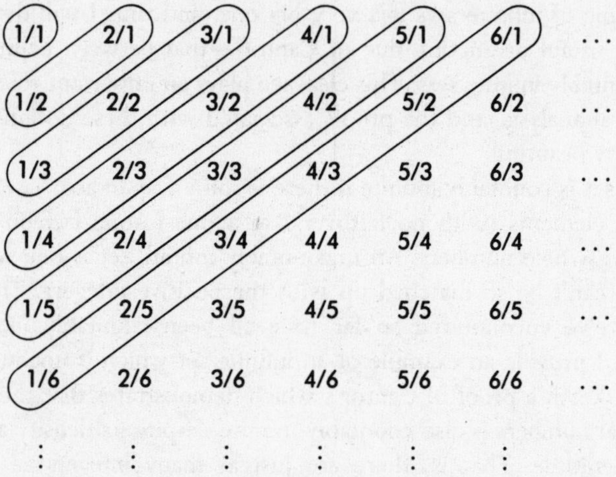

1/1 2/1 3/1 4/1 5/1 6/1 ...
1/2 2/2 3/2 4/2 5/2 6/2 ...
1/3 2/3 3/3 4/3 5/3 6/3 ...
1/4 2/4 3/4 4/4 5/4 6/4 ...
1/5 2/5 3/5 4/5 5/5 6/5 ...
1/6 2/6 3/6 4/6 5/6 6/6 ...
⋮ ⋮ ⋮ ⋮ ⋮ ⋮

Proof that the rational numbers are countable

Now for a set that is uncountable. Cantor showed that the set of all real numbers (all decimals) is more numerous (more infinite) than the set of integers or the set of rational numbers. More precisely put, there isn't any way to match up the real numbers with the integers (or the rationals) in a one-to-one fashion without there always being some real numbers left over. The standard proof of this fact is an indirect one and begins with the assumption that there is such a listing or match-up. Suppose, just to be specific, that the number 1 were matched with the real number 4.$\underline{5}$6733951 . . . , 2 with 189.3$\underline{1}$299008 . . . , 3 with .33$\underline{9}$33337 . . . , 4 with 23.543$\underline{7}$9802 . . . , 5 with .9896$\underline{2}$415 . . . , 6 with 6,219.31218$\underline{4}$62 . . . , and so on. How can I be sure that, no matter how it is continued, this infinite list (or any other list I might construct) is certain to leave out some real numbers?

For the answer, consider that number between 0 and 1 whose Nth decimal place is occupied by a digit 1 more than the underlined digit in the Nth decimal place of the Nth number on the list. (You may want to reread the previous sentence in a quiet corner.) Given the particular list above, the number I'm talking about begins .620835 . . . since its first digit 6 is 1 more than 5, 2 is 1 more than 1, 0 is 1 more than 9, and so on. The number doesn't appear anywhere on the list since by definition it differs from the first number on the list in at least the first decimal place, from the second number in at least the second decimal place, from the third in at least the third place, and from the Nth number on the list in at least the Nth decimal place.

1	4. 5 6 7 3 3 9 5 1 . . .
2	189. 3 1 2 9 9 0 0 8 . . .
3	. 3 3 9 3 3 3 3 7 . . .
4	23. 5 4 3 7 9 8 0 2 . . .
5	. 9 8 9 6 2 4 1 5 . . .
6	6,219. 3 1 2 1 8 4 6 2 . . .
⋮	⋮

The number beginning .620835 . . . appears nowhere on the above list since it differs from the Nth number on the list in at least the Nth decimal place. The real numbers are not countable.

That's it. QED. The proof is complete. No matter what infinite list of real numbers anyone presents to us, we can always, by a similar technique, construct a real number not on that list. We conclude that there is no way the real numbers can be paired up with the integers without leftovers. The set of real numbers is uncountably infinite and is said to have a higher (infinite) cardinality than does the set of integers or the set of rationals.

[There are even sets which have higher cardinalities than does the set of real numbers, examples being the set of all subsets of real numbers or the set of all functions of real numbers. In fact, a whole hierarchy of infinite cardinalities exists, starting with $\aleph_0$, Cantor's symbol for the cardinality of the integers ($\aleph$ is aleph, the first letter of the Hebrew alphabet). As I mentioned, however, the distinction between countable and uncountable sets is for most mathematicians the only one that counts.]

Cantor speculated that there was no subset of real numbers which was more numerous than the integers yet less numerous than the real numbers (whose cardinality, Cantor suggested, should therefore be indicated by the symbol $\aleph_1$). This speculation has come to be called the continuum hypothesis and has never been proved. One compelling reason is that, as Gödel and American mathematician Paul Cohen have shown, it is independent of the other axioms of set theory; both the continuum hypothesis and its negation are consistent with our present understanding of sets. A new, plausible axiom might decide the issue, but despite the attempts of many eminent logicians and set theorists (not to mention a quixotic effort by me involving "generic sets" which the margin of this page is too small to accommodate), it remains undiscovered.

Returning to our Hotel Infinity example after this exhausting journey, we note that if each of the hotel's countably infinite rooms had a countably infinite number of windows, the number of windows would still be only countably infinite. That is, there would be no more windows in the hotel than rooms. (The proof of this is similar to the proof that the rational numbers are countable.) Finally, to end on a shattering note, let's ignore the obvious physical constraints and imagine that these infinitely many windows are numbered and that at 11:59, windows 1 to 10 are smashed while window 1 is repaired. One half minute later, windows 11 to 20 are smashed and window 2 is repaired. One quarter minute later, windows 21 to 30 are smashed and window 3 is repaired. The progression is clear—one eighth of a minute later . . . The question is: How many of these windows are broken and how many repaired at 12:00? The answer is that they've all been repaired by 12:00.

Time to check out.

LIMITS

Take a circle that is one foot in diameter and inscribe in it an equilateral triangle. Now inside this triangle inscribe a circle and then inside this smaller circle inscribe a square. Inside the square inscribe a still smaller circle, inside of which you next inscribe a regular pentagon. Continue with these nested inscribings, alternating between a circle and a regular polygon whose sides increase by one with each iteration. It's clear that the area of the inscribed figures decreases with each repetition, but what is the ultimate area achieved by this sequence of figures? At first glance it appears that it should be zero, the process leading only to an isolated point. Remember, however, that as the number of sides of the polygons increases, they become more circular and after a while the process becomes, almost at least, one of placing a circle inside another circle with very little loss in area from one step to the next. In any case, your time is up. The limit of this procedure is a circle concentric with the original one and having a diameter of approximately one inch, 1/12th that of the original.

Many other problems in geometry lead to limits, and the notion is traditionally said to be the fundamental concept of calculus, the branch of mathematics dealing with change. With it we can make sense of an instantaneous rate of change, say of a satellite's position in space, or of the exact sum of a continuously changing quantity, perhaps the total force against a sloping dam. (See the entry on *calculus*.) Without such a notion we're forced to rely on approximations and averages. Limits capture formally our intuition of something tending toward or approaching an ultimate value and make clear the connection between ideal figures and infinity.

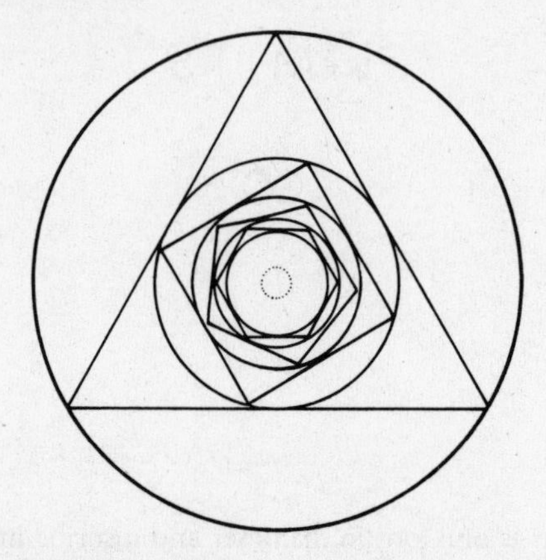

Nested inscribings which alternate between circles and regular polygons the number of whose sides increases by one with each iteration. The limit is a circle about 1/12th the diameter of the original.

Still, I question the central place limits are usually given in the beginning calculus course, where the primary consequence of their study is a drastic shrinkage of the pool of prospective mathematics, science, and engineering majors. Ignorance of the precise definitions of limit and its associated notions did not seriously hamper the inventors of calculus, Isaac Newton and Gottfried Wilhelm von Leibniz. They worked quite successfully with only an intuitive grasp of limits. In fact, utilizing calculus and his other seminal ideas on motion and gravitation, Newton helped usher in the most revolutionary developments ever in our conception of the physical world. Ignorance of such definitions didn't hinder the prolific work of Leonhard Euler and his countrymen the Swiss mathematical family Bernoulli in the eighteenth century. Nor, for that matter, does obliviousness to such notions impede many present-day physicists and engineers.

The nesting area problem above is not unusual in demanding some ingenuity but no more than a rough grasp of limits in order to solve.

The traditional judgment of the centrality of limits is true, however, in one important sense. Further theoretical developments in such mathematical specialties as differential equations, infinite series, the calculus of variations, and functional, real, and complex analysis all required a more rigorous foundation than that provided by Newton's "method of fluxions," which, taken literally, was nonsense.

This more rigorous foundation was laid in the nineteenth century by mathematicians Augustin Louis Cauchy, Richard Dedekind, and Karl Weierstrass. One part of it deals with the limit of a sequence of numbers which is defined to be the number L when, no matter how tiny a number ϵ is considered, the difference between the terms of the sequence and the number L eventually becomes (and remains) less than ϵ. Thus, in particular, 1/2, 3/4, 7/8, 15/16, 31/32 . . . approaches 1 since for *any* small number ϵ it can be shown that the difference between the terms in this sequence and the number 1 becomes (and remains) smaller than ϵ.

Once we have this definition (which is equivalent to several others) we may define the limit of a function $Y = f(X)$ as X approaches a number A. Formally, this limit is L if whenever a sequence of values for X approaches A as a limit, then the corresponding sequence of Y values (generated by the function) approaches L as a limit. Intuitively Y gets as close to L as we please whenever X is sufficiently close to A. With this definition in hand we can precisely characterize the central notion of the derivative of a function. By defining it as a limit (of a certain associated quotient function), we avoid many of the criticisms made against Newton. These complaints focused upon Sir Isaac's description of the instantaneous velocity of an object as the informal limit of quotients, distances traveled divided by times elapsed, and his subsequent need to explain how these vanishing quantities could both be equal to zero and not equal to zero. (If you're lost here, you're in good company. Don't worry.)

Although a *premature* emphasis on the many delicate points involving limits is wrongheaded, so is a total reliance on intuition. The precise definition of a limit is needed to clarify what is meant by the area of a curved region, the asymptotic limit of a curve or sequence of curves, the sum of a series $(1 + 1/3 + 1/9 + 1/27 + \ldots)$ (see the entry on *series*), and a host of other, more esoteric mathematical constructs. In an uncharacteristically gracious tone, Newton once remarked that if he had seen further than others it was because he had stood on the shoulders of giants. By standing on his and Cauchy's and a myriad of other shoulders, we see even further. The limit of this sequence of shoulder standings is indeterminate.

LINEAR PROGRAMMING

∞

Linear programming is a method of maximizing (or minimizing) some quantity while at the same time making sure that certain constraints on other quantities are satisfied. Since these constraints are generally linear (their graphs are straight lines), the discipline is called linear programming. It is one of the most useful techniques of operations research, the collection of mathematical tools developed after World War II to improve the performance of industrial, economic, and military systems, and has become a staple ingredient in the mathematics courses of business schools ever since. (See the entries on *Monte Carlo method of simulation* and *matrices.*)

Instead of invoking more wooden mathematical terms to clarify these, let's ponder a simple business break-even analysis for illustration. A small workshop produces metal chairs (or widgets, if you prefer more generic formulations). Its costs are $800 (for equipment, say) and $30 per chair produced. The total cost T incurred by the workshop is thus given by the formula $T = 30X + 800$, where X is the number of chairs produced. If we further assume that the selling price for these chairs is $50 apiece, then the total revenue R taken in by the workshop is given by the equation $R = 50X$, where X is the number of chairs sold.

Graphing both these equations on the same pair of axes, we find that they cross at that point at which the costs and the revenues balance. The break-even point is (40, \$2,000), meaning that if fewer than 40 chairs are sold, the cost exceeds the revenue; if more are sold, the revenue exceeds the cost; and if exactly 40 chairs are sold, the revenue and the cost are each \$2,000. Maximizing profit in this case is simply a matter of selling as many chairs as possible. (To obtain the break-even point 40 algebraically, subtract the equation $Y = 30X + 800$ from $Y = 50X$. The resulting equation is $0 = 20X - 800$, which yields $X = 40$.)

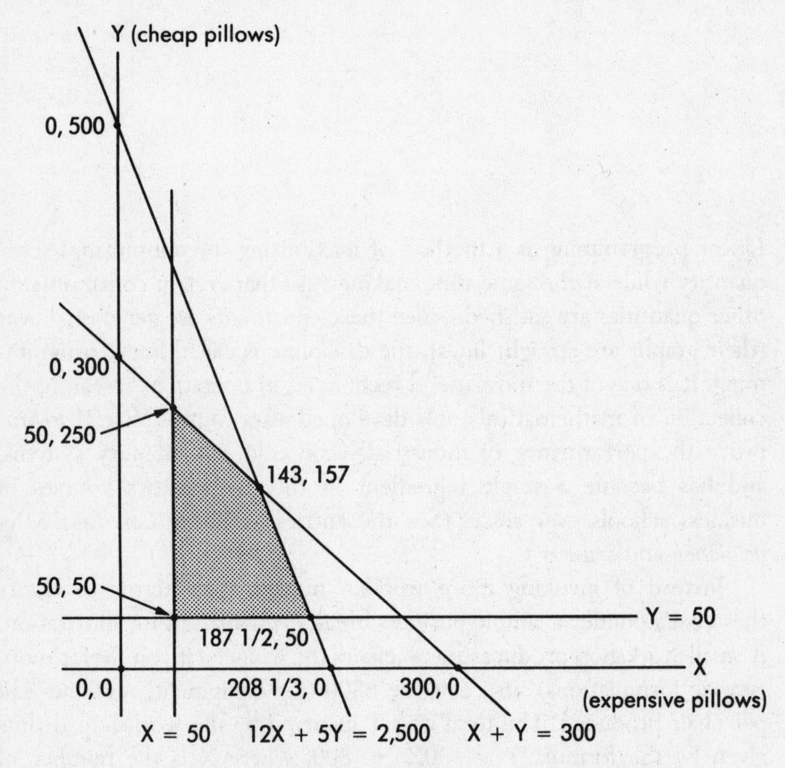

Shaded region satisfies all inequalities:
$X + Y \leq 300$; $12X + 5Y \leq 2{,}500$; $X \geq 50$; $Y \geq 50$.

With this as preparation consider the following genuine linear programming problem. Remaining with business applications, we assume that a company makes two kinds of pillows. The expensive pillows cost \$12 to make and sell for \$30, while the cheaper ones cost \$5 to make and sell for \$18. The company can make no more than 300 pillows per month and can spend no more than \$2,500 per month on their manufacture (featherbedding rules). If the company must make at least 50 of each type, how many of each should it make to maximize its profit?

Taking X to be the number of expensive pillows the company manufactures each month and Y the number of cheaper ones, we may translate the problem's restrictions on X and Y as: $X + Y \leq 300$; $X \geq 50$; $Y \geq 50$; and $12X + 5Y \leq 2{,}500$. The last inequality holds since each expensive pillow costs \$12 to manufacture and hence making X of them costs \$12X; likewise, making Y cheap ones costs \$5Y. Note that these contraints are expressed as linear inequalities, the graphs of which are regions bounded by straight lines (or, in more complicated problems, by their higher-dimensional analogues).

The quantity to be maximized is profit P, and in terms of X and Y, $P = 18X + 13Y$. This is so since each expensive pillow earns \$18 in profit (\$30 − \$12), each cheap one \$13 (\$18 − \$5), X of the former therefore earning \$18X, and Y of the latter earning \$13Y. Once the problem is set up in this manner, there are several techniques to determine its solution. One is graphical and depends on finding the extreme corners and edges of the feasible region—that part of the plane where all the inequalities hold true—and testing them to determine which yields the maximum profit. Using this method and a little analytic geometry, we discover that the pillow company should make 143 expensive and 157 cheap pillows each month if it wants to maximize its profit.

Another technique, called the simplex method and due to the American mathematician George Danzig, expands on and formalizes this geometric approach so that a computer can check these points rapidly when there are more than two variables. In use for more than forty years, the simplex method has saved an incalculable amount of money and time. Still, if the optimization problem contains many thousands of variables and linear inequalities, as is the case, for example, in the scheduling of airplanes or in the routing of telephone calls, then the checking may be a little slow even for a computer. In these situations a new algorithm developed recently by Narenda Karmarkar, a researcher at AT&T Bell Laboratories, is often quicker at finding the most efficient schedule or the shortest route.

When the constraints are not linear, the problems are much less tractable. I'm pleased to report that nonlinear programming problems frequently choke the largest supercomputers.

MATRICES AND VECTORS

∞

The original Latin meaning of the word "matrix" is womb or uterus. By extension it's come to mean that within which or from which something originates or develops. The mathematical definition is sterile by comparison: A matrix is a table of numbers arranged in rows and columns, the dimension given by two integers which indicate the number of rows and columns. The matrices $\begin{pmatrix} 2 & 2 & 3 & -1 & 9 \\ 8 & 3 & 7 & 11 & -2 \end{pmatrix}$ and $\begin{pmatrix} 7 & 6 \\ 1 & 0 \\ 3 & 6 \end{pmatrix}$ are 2 by 5 and 3 by 2 matrices, respectively, while a matrix having only 1 row and N columns is generally called an N-dimensional vector. These definitions are not very exciting. Tabular arrays of numbers have probably been familiar from the time of the earliest Phoenician CPAs. What's relatively modern are the ways in which this simple notational tool has been interpreted and the properties of the algebraic system which results when arithmetic operations are defined on matrices.

The most common use of matrices in mathematics involves the solving of collections of linear equations that arise in many physical and economic contexts. (See the entry on *linear programming*.) The method is similar to that used in elementary algebra. To simultaneously solve $12W + 3X - 4Y + 6Z = 13$, $2W - 3Y + 2Z = 5$, $34W - 19Z = 15$, and $5W + 2X + Y - 3Z = 11$ requires that one multiply various of these equations by numbers chosen so that when the equations are added or subtracted in pairs, the variables drop out successively. After solving a few systems of this sort, automating this tedious procedure becomes a very appealing idea, and one is led to the practice of performing these computations only on the matrix made from the equations' numbers (or coefficients) and leaving the equations themselves out of it. For the above system, the matrix of coefficients is $\begin{pmatrix} 12 & 3 & -4 & 6 & 13 \\ 2 & 0 & -3 & 2 & 5 \\ 34 & 0 & 0 & -19 & 15 \\ 5 & 2 & 1 & -3 & 11 \end{pmatrix}$, and the various arithmetic operations performed on the rows of this matrix (which correspond to equations) reduce it to a simpler matrix from which the equations' solution may be read.

In addition to these operations on the rows of a matrix, there are operations on the matrix as a whole that are essential for other applications. To keep computation to a minimum, let me first consider only the matrices A and B: $\begin{pmatrix} 3 & -6 \\ 5 & 2 \end{pmatrix}$ and $\begin{pmatrix} 1 & 0 \\ 3 & -8 \end{pmatrix}$, respectively. The matrices (A + B), (A − B), 5A, (5A − 2B), and A*B are, respectively: $\begin{pmatrix} 4 & -6 \\ 8 & -6 \end{pmatrix}$; $\begin{pmatrix} 2 & -6 \\ 2 & 10 \end{pmatrix}$; $\begin{pmatrix} 15 & -30 \\ 25 & 10 \end{pmatrix}$; $\begin{pmatrix} 13 & -30 \\ 19 & 26 \end{pmatrix}$; $\begin{pmatrix} -15 & 48 \\ 11 & -16 \end{pmatrix}$. Each entry (or component) in the sum, (A + B), is obtained by adding the corresponding entries in A and B. For (A − B), the corresponding entries are subtracted (if your algebra is rusty, recall that subtracting a −6 is equivalent to adding a +6). To multiply a number and a matrix, merely multiply each entry in the matrix by the number. This explains how both 5A and −2B are obtained. Putting the two together yields the matrix (5A − 2B).

And what about the product A*B? The entry in the first row and first column of A*B is [(3 × 1) + (−6) × (3)], or −15; it's computed by multiplying, componentwise, the entries in the first row of A by those in the first column of B and adding the results. The entry in the first row and second column of A*B is 48 and is computed by multiplying, again componentwise, the first row of A by the second column of B and adding the results. And, in general, the entry in the ith row and jth column of any matrix product is determined by multiplying, entry by corresponding entry, the ith row of the first matrix by the jth column of the second and summing the products. Applying this procedure, we get A*B = $\begin{pmatrix} -15 & 48 \\ 11 & -16 \end{pmatrix}$. Finally, the matrix I = $\begin{pmatrix} 1 & 0 \\ 0 & 1 \end{pmatrix}$

is called the identity matrix (womb for twins?) since C*I = I*C = C for all matrices C.

So what? One purely mathematical consequence of these definitions is that the set of matrices constitutes an algebraic structure called a noncommutative ring. I'll forgo the definition of a ring (roughly, it's a set with a couple of operations defined on its members and satisfying certain properties) and note that "noncommutative" means that, unlike the case with numbers, A*B needn't equal B*A. Here A*B is $\begin{pmatrix} -15 & 48 \\ 11 & -16 \end{pmatrix}$ but B*A is $\begin{pmatrix} 3 & -6 \\ -31 & -34 \end{pmatrix}$.

Why matrix multiplication is not commutative is made clearer by some understanding of vectors. So, continuing on my relentlessly didactic course in this entry, I declare that N-dimensional vectors, 1 by N matrices, are used to designate quantities that have not only a magnitude (such as temperatures or weights) but also a direction (such as forces or electromagnetic fields). It's often helpful, in fact, to think of vectors as arrows having the appropriate length and pointing in the appropriate direction.

Velocity is a typical vector quantity. A wind velocity of 10 miles per hour might be indicated by any of the vectors (0,10), (10,0), (0, −10), or (−10,0) depending upon whether the wind direction was, respectively, *to* the north, east, south, or west. In each case the first number is the component of the velocity in the east-west direction, the second the component in the north-south direction. The vector (−7.1,7.1) has a length of 10 [determined by the Pythagorean theorem: $(-7.1)^2 + 7.1^2 = 10^2$] and thus may also be thought of as indicating a velocity of 10 miles per hour, but in a direction between west and north (otherwise known as northwest). The vector (9.4,3.4) indicates a velocity of 10 as well (since $9.4^2 + 3.4^2 = 10^2$), but this time in a direction 20 degrees north of east. A wind whose velocity components were (28.2,10.2), three times as great as (9.4,3.4), would blow in this same direction but at 30 miles per hour.

More generally, vectors are used with quantities that require two or more dimensions to be specified and needn't represent anything physical. Five-dimensional vectors for restaurants might report restaurants' numerical ratings according to five different criteria.

However we interpret vectors, matrices may be viewed as representing transformations of them; a vector multiplied by a matrix (defined just as the product of two matrices) is transformed into another vector. The vector (1,1) for example, is transformed into $(8,-4)$ by $\begin{pmatrix} 3 & -6 \\ 5 & 2 \end{pmatrix}$ since $(1,1) * \begin{pmatrix} 3 & -6 \\ 5 & 2 \end{pmatrix} = (8,-4)$. These "linear transformations" stretch, rotate, and reflect vectors (although if the vectors do not denote physical quantities, these stretchings, rotations, and reflections are only figures of speech); they can always be represented by matrices.

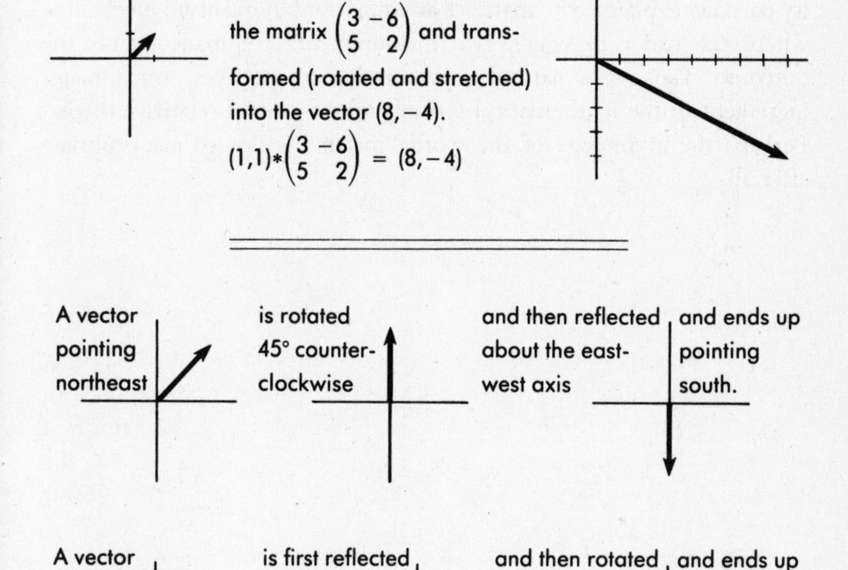

These transformations do not commute, and thus neither do the matrices representing them.

Performed in succession, these transformations of vectors needn't be commutative. For example, a rotation of a vector followed by a reflection of the resultant vector isn't always the same thing as a reflection followed by a rotation. (To see this, imagine a vector pointing northeast being rotated 45 degrees counterclockwise so that it's pointing north. If this vector is now reflected about the east-west axis, it will end up pointing south. If instead the reflection of the vector about the east-west line occurred first, followed by a counterclockwise rotation of 45 degrees, the vector would end up pointing east.) Thus the matrices representing these rotations, reflections, and other transformations needn't commute either; A*B doesn't always equal B*A.

Matrices and vectors play a primary role in linear algebra as well as in many other areas of applied mathematics. The failure of commutativity partially explains why matrices are important in quantum mechanics, where the order in which two measurements are made affects the outcome. Tensors, a natural generalization of matrices, are a major ingredient in the mathematical formulation of general relativity theory. Perhaps the etymology of the word "matrix" is not so inappropriate after all.

MEAN, MEDIAN, AND MODE

∞

The fourth-grader notes that half the adults in the world are men and half are women and concludes therefrom that the average adult has one breast and one testicle. A real estate agent informs you that the mean price for a house in a certain neighborhood is $400,000 and implies that there are many houses over that price in the neighborhood. A salesman tells you that the median commission on the nine sales he made that day is $80 and suggests that he therefore made $720 on these sales. The caterer says that the mode, or most common, lunch spread for his parties runs $1,200 and insinuates that half his customers spend more. A stockbroker claims your investment will be worth millions of dollars, but your calculations indicate hundreds is more like it.

Of the five we can be sure only of the fourth-grader's claim. The mean, median, and mode are all "middle indicators" or measures of central tendency, numbers that are supposed to give one a feeling for what is typical or standard in a given situation, but don't always do so. Since their relative values can vary considerably, it's important to know the definitions. (See also the entry on *statistics—two theorems.*)

The mean of a set of numbers is what is usually called the average (or the arithmetic average) of these numbers. To find the mean of N numbers, you merely add the numbers together and then divide by N. The definition is easy and familiar, but some of the inferences people draw from it are unfounded. For example, the neighborhood referred to above may contain very few houses in excess of $400,000; perhaps it has a few large mansions engulfed by a dreary swarm of quite modest tract houses.

By contrast, the median of a set of numbers is the middle number in the set. To find it, merely arrange the numbers in increasing order; the median is the middle number (or the average of the two middle numbers if there are an even number of numbers in the set). Thus, the median of the set 8, 23, 9, 23, 3, 57, 19, 34, 12, 11, 18, 95, and 48 is found by listing them as 3, 8, 9, 11, 12, 18, 19, 23, 23, 34, 48, 57, and 95 and then noting that 19 is in the middle and is hence the median for this small set of numbers, whose mean, incidentally, can be computed to be about 27.7. (In a very large collection of numbers, the median is sometimes referred to as the 50th percentile, indicating that it is greater than 50% of the numbers in the collection. Likewise, a number at the 93rd percentile is assumed to be greater than 93% of the numbers.)

The salesman mentioned above could very well have earned thousands of dollars in commission on his nine sales which had a median value of $80; maybe six of the sales netted him $80 apiece, while the remaining three netted him $5,000 each. His median commission would then be $80 as he said, but his total commission would be $15,480, much more than the $720 he suggested he earned. His mean commission in this case would be $1,720.

At times an even more misleading figure, the mode of a set of numbers is simply the number which appears most frequently and needn't be anywhere near the mean or the median of the set. The caterer who cited the $1,200 figure as the mode lunch spread might have had the following orders for the month: $400, $800, $800, $1,200, $800, $1,200, $200, $1,200, $200, $400, $200, $1,200, $400. The mode is certainly $1,200, but the median is $800 and the mean is only $692.

A somewhat more sophisticated and tricky example of the difference between the mean and the mode of a quantity is provided by the man who invests $1,000 in a volatile stock that each year with equal probability either rises by 60% or falls by 40%. He stipulates that the stock is to be sold by his great-granddaughter in 100 years and wonders how much she will receive. The amount depends upon the number of years the stock will have risen, but its mathematical mean, also called its expected value in probabilistic contexts, is a whopping $13,780,000. The mode or most likely value of her inheritance is, however, only a paltry $130.

The explanation for this disparity is that the astronomical returns associated with many years of 60% growth skew the mean upward, while the tiny returns associated with many years of 40% shrinkage are bounded below by $0. The problem is a contemporary version of the so-called St. Petersburg paradox. [For those who want more detail: The stock rises an average of 10% each year (the average of +60% and −40%). Thus after 100 years, the mean or expected value of the investment is $\$1{,}000 \times (1.10)^{100}$, which is $13,780,000. On the other hand, the most likely result is that the stock will rise in exactly 50 of the 100 years. Hence the mode is $\$1{,}000 \times (1.6)^{50} \times (.6)^{50}$, which is $130. The mathematical expected value is not always the value expected.]

More commonly the expected value of a quantity is calculated by multiplying the possible values of the quantity by the probabilities that these values will be attained and summing these products. Consider as an illustration a home insurance company which has good reason to believe that each year, on average, one out of every 10,000 of its policies will result in a claim of $400,000; one out of 1,000 policies will result in a claim of $50,000; one out of 50 will result in a claim of $2,000; and the remainder will be no problem. The insurance company wants to know what its average payout per policy is (to know what premiums it might charge), and the answer is the expected value—in this case: ($400,000 × 1/10,000) + ($50,000 × 1/1,000) + ($2,000 × 1/50) + ($0 × 9,789/10,000) = $40 + $50 + $40 + $0 = $130.

Understanding these various measures of central tendency will make the average person 36.17% less likely to fall victim to misleading uses of these figures by real estate brokers, salesmen, stockbrokers, and caterers. Of course, this same average person has already fallen victim to a severe case of hermaphroditism, so she/he probably will have other worries.

MÖBIUS STRIPS AND ORIENTABILITY

∞

Take a small can of tuna and carefully remove the label. The long rectangular strip of paper will have writing on one side and be blank on the other. Give this strip of paper a half twist and tape the two ends together, making sure that the blank side meets up smoothly with the outer, written-on part. What you have is a Möbius strip, and it is known for the strange topological properties it possesses.

Primary among these properties is the fact that the Möbius strip has only one side. There is a continuous change from blank to writing to blank. Alternatively, I can declare that you would not be able to collect from someone who promised you $1,000,000 if you could paint one side of the Möbius strip red and the other side blue. Starting with red anywhere on the strip and painting steadily, you would finally come back to where you began and the whole strip would be red.

To understand another odd property of this figure, it's helpful to imagine a line along its middle. Cutting along this line all the way around the Möbius strip would, it seems obvious, separate it into two pieces. But no. What results is simply a longer strip. Or consider instead cutting a Möbius strip along a line parallel to an edge, but this time starting at a point a third rather than a half of the strip's width away from the edge. What results in this case is two interlinked strips, one of them a Möbius strip.

The one-sided Möbius strip is one of the best known of a large number of topological aberrations. (See the entry on *topology.*) Although it itself has no important applications or manifestations in nature (so far at least), its counter-intuitive simplicity is appealing. Given this simplicity, it seems remarkable that this little half twist wasn't discovered earlier, but the honor of bringing it into the world belongs to A. F. Möbius, a nineteenth-century German astronomer.

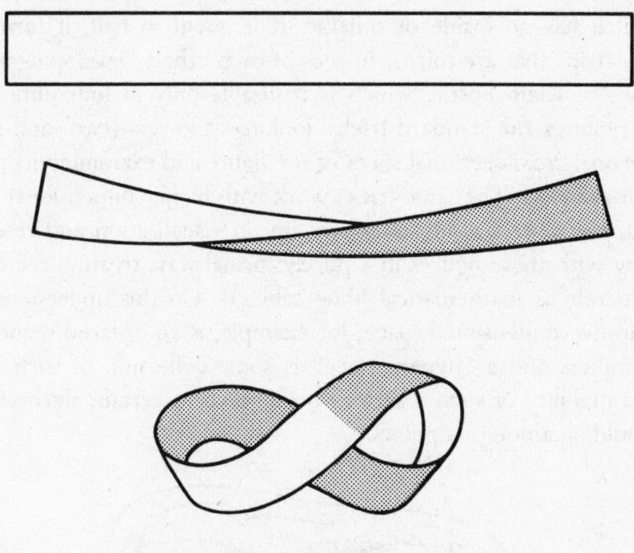

The Möbius strip

Möbius also noted that no consistent way of assigning orientations on a Möbius strip is possible. To see this, imagine sliding a two-dimensional hand-shaped object once around a Möbius strip. Remembering that the ideal Möbius strip has no thickness and that therefore the hand is visible from both "sides" of the strip, we observe that the hand's orientation is reversed when it returns to its starting point. A left hand becomes a right hand and vice versa. Physicists have speculated that if the universe were "nonorientable" like the Möbius strip (cosmically dyslexic, if you like), an astronaut making a long trip could conceivably come back to earth with her heart on the right side of her chest.

The notion of orientation or handedness is dependent upon that of dimension. If you cut out two similar hand-shaped pieces of cardboard, one left and one right, and slide them around on the floor, there is no way they can be made coincident. But if you lift one of the "hands" up into the third dimension and simply flip it over onto the other one, this can be done. The Möbius strip also accomplishes this flip through the third dimension, but in a more twisted manner reflecting the peculiar way in which it's embedded in three-dimensional space.

A three-dimensional analogue of the Möbius strip is the Klein bottle, which has no inside or outside. If it is cut in half, it forms two Möbius strips that are mirror images of each other! Developing a visual feel for the Klein bottle, which is realizable only in four-dimensional space, requires the standard tricks: looking at lower- (two- and three-) dimensional cross-sectional slices of the figure and examining its projections or shadows. The same tricks work with higher-dimensional figures as well, but after a while one gives up on visualization and resorts to working with these figures in a purely formal way, treating the dimensions merely as mathematical filing cabinets. On this understanding, a point in five-dimensional space, for example, is an ordered sequence of five numbers and a "hypersurface" is some collection of such points. Nonorientability of such a surface boils down to certain algebraic relations holding among its points.

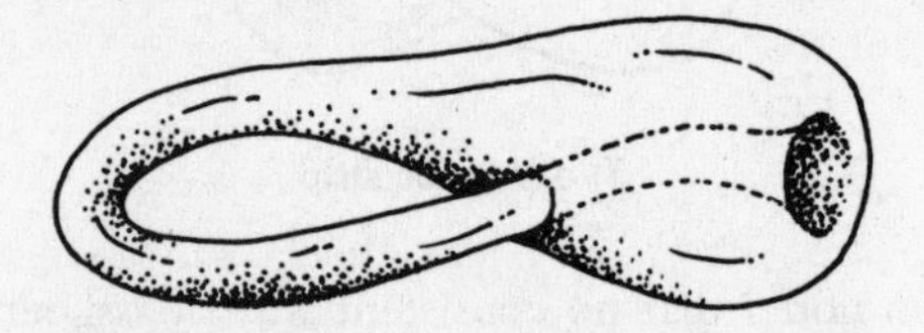

A Klein bottle has no inside or outside. It is realizable only in four-dimensional space, where it doesn't intersect itself, as it seems to do in this two-dimensional depiction.

MONTE CARLO METHOD OF SIMULATION

∞

A basketball player is known to hit 40% of his shots. He takes 20 shots in a game. What is the probability he will make exactly 11 of them? There are certain standard calculations you can make that will yield the answer. There is also another method which, optional in this case, is sometimes the only way to attack a problem. What it amounts to here is asking the basketball player to play a quick 10,000 games so that we can find the percentage of times he makes exactly 11 of his shots.

Clearly not feasible for a human basketball player, this so-called Monte Carlo method of simulation can easily be implemented on a computer. Simply ask the computer to generate a random whole number between 1 and 5, and note if this number is a 1 or a 2. Since 2 is 40% of 5, obtaining a 1 or a 2 we will understand to be the simulation of the player's making a basket, while obtaining a 3, 4, or 5 we will take to be the missing of a shot. Next ask the computer to generate 20 such random numbers between 1 and 5 and note if exactly 11 of them are either 1's or 2's. Should this occur, we'll understand it to be the equivalent of the player's making exactly 11 baskets in 20 tries when his shooting percentage is 40%. Finally, ask the computer to perform this little exercise 10,000 times and to keep track of the number of times 11 of the 20 attempts per game result in simulated baskets. Divide this number by 10,000 to get a very good approximation to the theoretical probability in question.

To get a feel for simulation it's helpful to actually perform one yourself. (Don't worry. You don't need a computer; a penny will do.) Imagine that you're called in as a statistical consultant by a country's sexist government. It has just adopted a policy requiring couples to continue having children until they have a boy, after which time they must stop. The questions the country's rulers want answered are: How many children will the average family have under this policy and what will the distribution of the sexes be? Instead of collecting statistics that will of necessity take years to amass, you can flip coins to come up with a large enough sample upon which to make an estimate. Interpreting a head (H) to be a boy and a tail (T) to be a girl, you flip a coin until a head turns up and then record the number of flips—i.e., the number of children in the family. The sequence TTH corresponds to two girls followed by a boy, H corresponds to an only boy child, and so on. Repeat this procedure 100 or 1,000 times to produce 100 or 1,000 "families" and compute the average number of children per family and the sex distribution. You as well as the country's officials may find the answers surprising.

Studies of large systems, queuing and scheduling situations, physical, engineering, and mathematical phenomena are all facilitated by the use of such Monte Carlo methods. From retail chain stores to aeronautical turbulence laboratories, everybody's simulating. The "what if" capabilities of business spreadsheets are a form of economic simulation. Generating random numbers on a computer and then manipulating the probabilistic simulations that are based on them is easier and less expensive than dealing with real-life chance phenomena. The only caveat is that one remember that there is a sharp difference between a model or a simulation of a phenomenon and the actual event itself. Having a baby is not the same as flipping a coin.

The following schematic representation of simulation—and, to an extent, of applied mathematics in general—is helpful in this regard. The process may be divided into five stages: The first is identifying the raw phenomenon in which you're interested; the next, creating an idealized version of this phenomenon; the third, constructing a mathematical model based on this simplified version; then, performing various mathematical operations on the model in order to obtain predictions; and last, comparing these predictions with the original raw phenomenon to see if they make sense. (See the entry on the *philosophy of mathematics.*)

Many applications of mathematics are straightforward, but it's distressingly easy, especially in the social sciences, to identify one's model with "reality" and to impute some property to reality that exists only in the model. One simple example from elementary algebra: George can perform some task in 2 hours, Martha the same task in 3 hours. How long will it take them working together? The "correct" answer of 1 hour and 12 minutes assumes that they work seamlessly together, neither hindering nor enhancing the other's effort. In this case and in countless others, the certainty of the mathematical conclusions derived from the model does not always extend to the assumptions, simplifications, and data that one uses to construct it. The latter are sticky, nebulous, and quite fallible despite the sometimes annoyingly smug claims of sociologists, psychologists, and economists. Reality, like Virginia Woolf's perfectly ordinary woman, Mrs. Brown, is indefinitely complex and impossible to capture completely in any mathematical model.

[The answer to the simulation problem is that the average number of children per family is two, one boy and one girl.]

THE MULTIPLICATION PRINCIPLE

The St. Ives Mother Goose rhyme poses a problem that appears in almost identical form in the ancient Egyptian Rhind papyrus dating from 1650 B.C.: "As I was going to St. Ives, I met a man with seven wives. Every wife had seven sacks, and every sack had seven cats, every cat had seven kittens. Kittens, cats, sacks and wives, how many were going to St. Ives?" The answer is one, since everyone else is heading away from St. Ives, but determining how large a group is encountered depends on an understanding of the multiplication principle.

No other idea in combinatorial mathematics is as simple, yet as far-reaching as this bland-sounding principle: If one can perform some action or make some choice in one of M different ways and if afterward one can perform another action or make another choice in one of N different ways, then one can perform these actions in succession or make these choices in succession in M × N different ways.

Assume, for a less ancient example, that you find yourself stranded in Los Angeles and need desperately to get to the East Coast; anywhere on the East Coast will do. You consult the airline schedules and discover there are no direct flights, but there are flights to three midwestern cities (say Minneapolis, Chicago, and St. Louis), from each one of which there are flights to four East Coast cities (say Boston, New York, Philadelphia, and Washington). The multiplication principle tells you that there are 12 (= 3 × 4) different ways to reach the Atlantic Coast.

These ways can be symbolized as MB, MN, MP, MW, CB, CN, CP, CW, SB, SN, SP, and SW, where the letters refer to the cities mentioned.

The principle can be applied more than once. For example, if a school's enrollment is 18,000, we can be absolutely certain that at least two students at the school have the same three initials. The reason is that the 26 possible first initials may be followed by any of 26 possible middle initials which may in turn be followed by any of 26 possible last initials. Thus by the multiplication principle, there are 26^3 or 17,576 sets of three (ordered) initials. This number is smaller than the number of students enrolled, so we conclude that at least two students must have the same set of three initials. (In fact, at least 424 students share initials, and as a practical matter it's probable that many more do.)

Braille symbols, which all consist of two vertical columns of three dots each, provide another example. The different letters and symbols in this alphabet for the blind are distinguished by the fact that different subsets of the six dots are raised. For example, the letter "a" is indicated by only the upper-left-hand dot being raised, whereas the letter "r" is indicated by the three dots in the left column as well as the middle dot in the right column being raised. How many symbols are there in all? For each dot, we have two possibilities—to raise it or not. Thus since there are six dots, there are 2^6 or 64 different possibilities. One of these 64 has no raised dots and is hence imperceptible, so there are 63 different Braille symbols (letters, numbers, letter combinations, common words, and punctuation symbols).

Or consider a coded message in English that must take the form SPOOK7, where the first two entries must be consonants, the next two entries vowels, the next a consonant, and the last a number between 1 and 9. There are ($21^2 \times 5^2 \times 21 \times 9$) or 2,083,725 possible messages of this form. If all the symbols in the message must differ, there are only ($21 \times 20 \times 5 \times 4 \times 19 \times 9$) or 1,436,400 such messages. The number of telephone numbers possible within an area code is approximately 8×10^6 or 8 million, since the first number may be any of the eight digits other than 0 or 1, and the subsequent six numbers may be any of the 10 digits. (In practice there are more constraints on allowable phone numbers, but we'll ignore them here.) If a state's license plates have the pattern L L L L N N, four letters followed by two digits, then that state has ($26^4 \times 10^2$) or 45,697,600 possible license plates. If the letters and digits must all differ, there are only ($26 \times 25 \times 24 \times 23 \times 10 \times 9$) or 32,292,000 different plates.

The numbers that department stores, utilities, and credit card companies use to identify us are often 15 or 20 symbols long, far longer than they need be, given the population of the United States. Even if they're comprised of digits alone, 20-symbol sequences are sufficient to assign an ID to 10^{20} people, 20 billion times the population of the whole world. One reason for this is that the extra capacity makes it very unlikely that would-be impostors or crooks will be able to come up with a sequence which belongs to a paying customer.

As these examples attest, a surprising consequence of the multiplication principle is the large number of possibilities that result from its repeated application. This number grows exponentially with the number of times the principle is applied, and even the fastest computers, if they attempt to enumerate possibilities or use other brute force methods to solve large problems, soon run up against what is called the combinatorial explosion and grind to a crawl.

The same problem of burgeoning possibilities (but on a much more modest scale) plagues the occasional efforts of authors or directors to have a number of junctures in their books or movies where the reader or viewer may express a choice about how to proceed. If there were only 5 such junctures, 32 different books or movies would have to be made to accommodate these choices. (There would be two choices at the first juncture, each branch of which would lead to a juncture, which would in turn lead to two branches each having a juncture, and so on, resulting in 2^5 or 32 different treatments.) If there were more junctures or more choices at each juncture, the number would be vastly larger. In fact, the number of works necessary to replicate the feeling of freedom inherent in even a brief conversation with someone would require an author or director to devote a lifetime to all possible depictions. (See the entry on *human consciousness*.)

This idea of continuous branching underlies the many metaphors we have about friends drifting apart, histories diverging, and people growing eccentric with age. Also, as hinted, it plays a significant role in our conception of freedom and (for a more outré example) in the so-called many-worlds interpretation of quantum mechanics in which the universe splits at every instant into an infinity of noncommunicating universes.

The multiplication principle has many other applications and variants, the most useful and best-known of which involve the combinatorial coefficients. (See the entry on *Pascal's triangle* for a discussion of these.)

[The size of the group in the Mother Goose rhyme is 2,801. What would the answer be if each kitten carried seven fleas?]

MUSIC, ART, AND DIGITALIZATION

∞

Music and number have been linked from the time of Pythagoras and his followers. They were the first to point out the connections between mathematical ratios and harmonious sounds. Taut strings whose lengths were in whole-number ratios brought forth euphonious tones when plucked, and the extension of string lengths in whole-number ratios produced an entire scale.

Skipping ignorantly over two millennia, I note that the work of early-nineteenth-century mathematician Joseph Fourier enlarged considerably on this knowledge and enabled people to mathematically describe arbitrary musical sounds as combinations of periodic trigonometric functions. A sound's pitch, loudness, and quality could be related, respectively, to the frequency, amplitude, and shape of the periodic functions representing it.

Modern electronic music builds on these and other insights. From the computer compositions of John Cage to the latest advances in recording studio techniques, mathematics is intimately involved in music and music processing.

Much less mathematical than music, art displays few parallels in its development. The rough analogue of Pythagorean ratios might perhaps be the use of projective geometry and perspective by Albrecht Dürer, Leonardo da Vinci, and other Renaissance artists, and the evolution of computer graphics and fractal geometry may bear at least a superficial resemblance to digital music.

But this last adjective, "digital," hints at the reason the association between music and mathematics has been more natural than that between art and mathematics. Music is or can easily be made digital or discrete; at least until very recently art could not be. Observe, for example, that in musical notation as in number notation, position is important; where the note is on the staff determines its pitch. In fact, from the writing of notes and scores to the use of symmetries and recursion in works as diverse as Bach's and Cage's to the sound technology of organ construction and compact disks, numbers and mathematics have always played a significant supporting role in music. In contrast, "painting by the numbers" denotes vulgarity, as does the phrase "art processing."

This difference can be neatly expressed in terms of computers. Those early machines whose output varied gradually with a continuously changing physical quantity such as voltage or pressure were termed analog computers. This was in contrast to today's more familiar digital ones, whose output is all or none depending upon whether a particular logico/electronic condition is met or not. As even the above very schematic remarks on music and art suggest, however, the distinction crumbles when pushed too hard. A vibrating drumhead leads more naturally to an analog device than a digital device, and the editing of graphics on a pixel-by-pixel basis is as digital a procedure as there is. (A pixel is the smallest dot of light that a screen can show; most have something like 200 × 300 or 60,000 pixels.)

Still, the contrast is a useful one to make, and some version of it is apparent in a variety of contexts. Speedometers, for example, may be digital and give a numerical readout, or analog and produce a lengthening rectangle (or rotating pointer). The analog output is less precise, but puts one's speed in some perspective. An 82 and a 28 are almost indistinguishable, but a long rectangle is quite distinct from a short one, and if its length is changing, that too is manifest. The same trade-off is present with clocks and watches, digital timepieces producing precision, but lacking in all the associations a clockface can engender.

I've noticed a similar difference in the impact a piece of writing has on me depending upon whether I read it off a computer monitor or read a hard (i.e., paper) copy whose pages I can rustle through. The latter analogue of an analog display again provides more ambience than does the corresponding digital display on the screen. I get a better sense of the structure, heft, and proportions of the piece if I can handle the manuscript. Likewise, although very much in favor of the increased use of digital calculators in the classroom, I think that calculator answers should be checked against common sense and that an ability to estimate and compare magnitudes should be developed.

The digitalization of formerly analog devices often carries with it an aura of artificiality. The sound of instruments and voices can be easily modified and even fabricated by sound engineers and mixers, whose product, the compact disk, contains music never before heard in any natural environment. Much more surprising is the fact that photographs, which heretofore have been seen as providing documentary evidence of reality, can be altered by similar, but more sophisticated digital techniques. A picture of a man and a woman standing against a brick wall, for example, no longer means that that man and that woman ever did stand together against that wall. Photographs can be atomized and reconstituted electronically, so that colors, fabrics, surfaces, clothes, and faces can be altered at will, and the result will be indistinguishable from a "real" picture.

I don't want to put too much weight on this somewhat nebulous distinction between digital and analog, but it is also relevant to cognitive style. It persists in the computer world, for example, in the opposition between keyboard-based digital commands (which I prefer) and mouse-based ones, the latter having a more analog flavor to them. In mathematics too I usually prefer information digitally expressed in terms of words and symbols to analog expressions involving pictures and diagrams. Regardless of personal preferences, however, the gap between mathematical ways of communicating and verbal ones is smaller than most people realize, and thus I find it surprising that mathematics books seldom exploit the prodigious facility we all, numerate and innumerate alike, have with language. Why, since mathematical ideas are almost always clearly expressible in words, is fuller use not made of this most encompassing and powerful of tools?

Too often pictorial accompaniments are distracting anyway. I have a friend who has become so mesmerized by his computer graphics software that he can't write a personal letter without cluttering it all up with clip art and frequency polygons. (Admittedly this is done with words too, sometimes in annoying little parenthetical asides like this.) These geometric figures not only divert attention from the matter at hand but also intimidate many people with literary sensibilities and/or poor mathematical backgrounds. Needless to say, these biases account for the relative paucity of illustrations in this book.

I should stop here since I'm coming perilously close to sounding like those people who explain everything from the national debt to Icelandic epic poetry by mumbling some bromide about left brains and right brains. It suffices to says that music, art, and mathematics may make use of a different mix of skills and talents, but they all require whole brains.

NON-EUCLIDEAN GEOMETRY

∞

In the welter of facts about triangles and parallelograms, similarity and congruence, areas and perimeters, the deductive character of geometry is sometimes forgotten. Originating in Egypt and ancient Babylonia with practical rules of thumb for land surveying, commerce, and architecture, all these facts and rules were shown by the Greeks to follow from only a few of them. The basic idea is easy to state. One chooses some "self-evident" geometric assumptions called axioms and then derives from them, by logic alone, certain other geometric statements called theorems. In his development of the subject, Euclid chose five axioms (actually ten, but only five of them were geometric) and derived the beautiful and influential body of theorems known as Euclidean geometry. (See the entry on *the Pythagorean theorem.*)

One of Euclid's five axioms was the so-called parallel postulate. This axiom stated (and still states) that, through a point not on a given straight line, we can draw exactly one straight line parallel to the given line. A familiar consequence of the parallel postulate is the theorem that the sum of the angles of any triangle is always 180 degrees.

Since it didn't seem quite as intuitive as the others, mathematicians tried sporadically throughout the centuries to prove the parallel postulate using the other four axioms. They used every method imaginable but could never come up with a proof. This failure and the naturalness of the other axioms seemed to give Euclidean geometry a certain absoluteness. It reigned for a couple of millennia as a monument to both common sense and eternal truth. Immanuel Kant even claimed that people could think about space only in Euclidean terms. Finally, in the nineteenth century the mathematicians János Bolyai, Nikolai Lobachevski, and Karl Friedrich Gauss realized that Euclid's parallel postulate was independent of the other four axioms and could never be derived from them. Furthermore, they understood that one could add the negation of the parallel postulate to the other axioms and still have a consistent system of geometry.

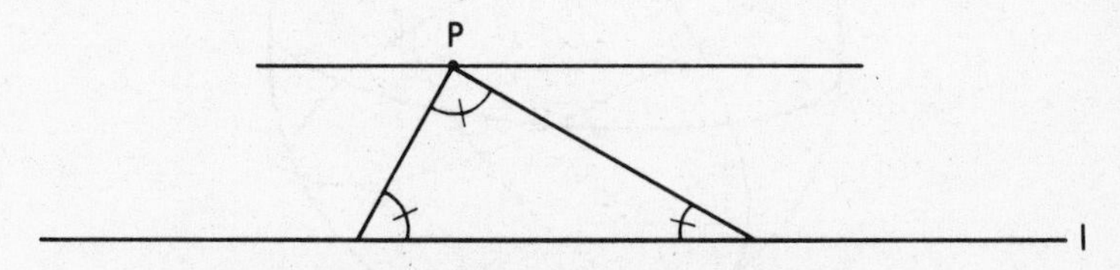

The sum of the angles of a triangle is 180°. There is exactly one line through P and parallel to l.

To see this, note that it is perfectly possible to interpret the basic terms of geometry in a wholly different way, yet still be bound by the strictest logic. Just as "all A's are B's, and C is an A, so C is a B" provides the warrant for many disparate arguments depending on the interpretations of A, B, and C, so the terms of Euclidean geometry may be interpreted quite unconventionally and still lead to valid theorems. For example, instead of the usual interpretations we may take "plane" to be the surface of a sphere, "points" to be points on the sphere, and "straight lines" to be the great circles of the sphere (any circular arc around the sphere which cuts it in half). If we do adopt these meanings, "straight lines" are still the shortest distance between two points (on the surface of the sphere) and we obtain an interpretation of geometry which makes true all of Euclid's axioms except the parallel postulate. Also true are all the theorems proved from these four axioms.

Verifying the axioms, we observe that between any two "points" there is a "straight line" since there is a great circle connecting any two points on the surface of the sphere. (Note that the great circle through Los Angeles and Jerusalem takes one over Greenland and is the shortest distance between the two cities.) Given any "point" and any distance, there is a "circle" on the surface of the sphere with that point as center and that distance as radius (just a normal circle on the sphere's surface). Also, any two "right angles" are equal, and any "line segment" (part of a great circle) may be indefinitely extended.

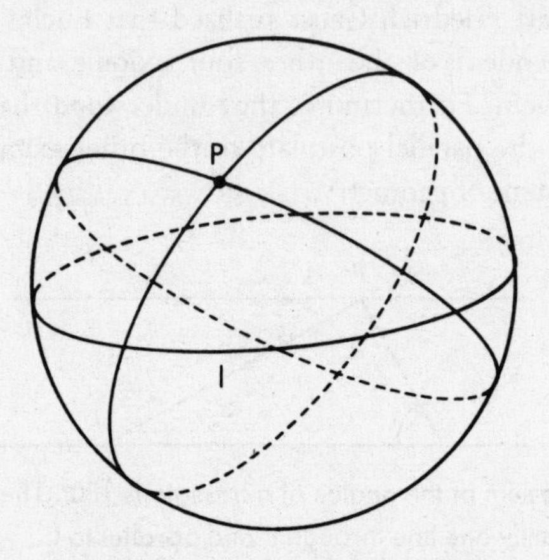

There is no "straight line" through P and parallel to the "straight line" l. All other Euclidean axioms are true under this interpretation.

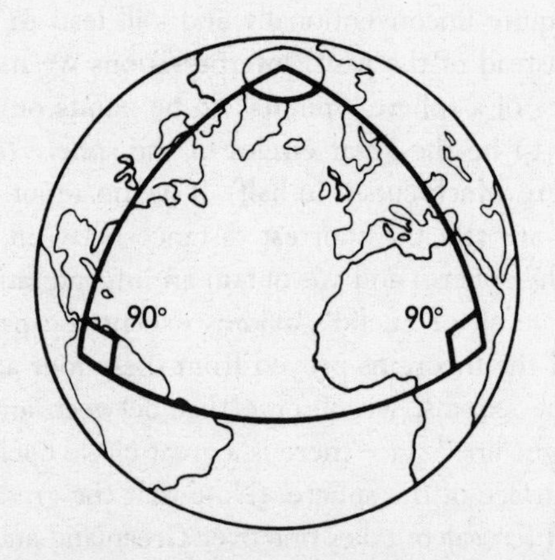

"Line segments" connecting Kenya, Ecuador, and the North Pole form a "triangle" whose angles sum to more than 180°.

The parallel axiom, however, does not hold for this particular interpretation of the terms, since given a "straight line" and a "point" not on it, there is no "straight line" parallel to the given line through the given point. As an illustration, take the equator as the "straight line" and the White House in Washington as a "point" not on it. Any "straight line" through the White House must be a great circle cutting the earth in two and, as such, will necessarily cross the equator and hence not be parallel to it. Another anomaly of this interpretation is that the sum of the angles of a triangle is always larger than 180 degrees. Summing the angles of the "triangle" formed by the part of the equator between Kenya and Ecuador and the "line segments" connecting "points" in these two countries with the North Pole will demonstrate why this is so. The spherical triangle thus formed has two right angles at the equator.

There are other nonstandard interpretations of the terms "point," "line," and "distance" in which all the axioms of Euclidean geometry hold true but the parallel postulate fails for a different reason: There is *more* than one line parallel to a given line through a point. In these models (which might, for example, be saddle-shaped surfaces) the sum of the angles of a triangle is less than 180 degrees. The German mathematician Bernhard Riemann conceived of still more general surfaces and geometries in which the concept of distance varies from point to point in somewhat the same way it does for a traveler along a very irregular and hilly terrain.

Any model in which the parallel postulate fails for whatever reason is said to be a model of non-Euclidean geometry. Each of the geometries discussed is a consistent set of propositions (just as the constitutions of various nations are differing but consistent sets of laws). Which of them is true of the real world depends on what physical meaning we give to the terms "point" and "straight line" and is an empirical question to be decided by observation, not armchair proclamations. Locally, at least, space seems to be as Euclidean as an Iowa cornfield, but as flat-earthers the world round have discovered, it's dangerous to extrapolate one's parochialisms too far. Taking the path of a light ray as the interpretation of a straight line results in a non-Euclidean physical geometry.

Finally, I like to think of the discovery of non-Euclidean geometry as a sort of mathematical joke—a joke that Kant did not get. Many riddles and jokes have the form "What has this, that, and the other property?" On hearing them, the thought that springs naturally to mind is quite different from the punch line's unexpected interpretation of the joke's conditions. Such is the case with non-Euclidean geometry. Instead of "What's black and white and red all over?" we have "What satisfies Euclid's first four axioms?" The new punch line was supplied by Bolyai, Lobachevski, and Gauss: stand-up comics at Club Universe.

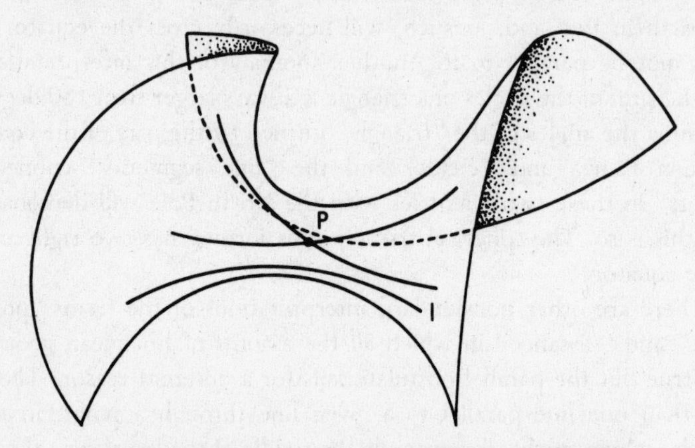

If on this saddle-shaped surface we interpret "straight line" to be the shortest distance between points, all the axioms of Euclidean geometry hold true except for the parallel postulate. Through P there passes more than one line parallel to l.

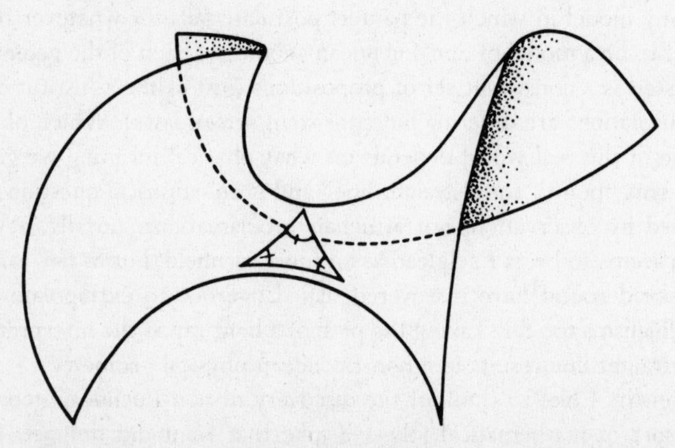

The angles of a triangle on this surface sum to less than 180°.

NOTATION

∞

There are six things to remember about good mathematical notation, and I'll refer to them by 1, b, III, four, E, and vi, respectively. The above list, adapted from a joke by comedian George Carlin, exemplifies a particularly bad choice of notation. Except for extreme cases such as this, however, most people, if they think about it at all, believe that mathematical notation can cause no confusion and lead to no insight, that it is rather a superficial, almost cosmetic aspect of mathematics. How we denote lines, angles, points, numbers, and other mathematical entities: What a fabulously trivial and unimportant matter that is. The pedant's perennial preoccupation with notation only strengthens this natural belief, but it is not always well founded. Systems of notation can sometimes be more than grandiloquent typologies and empty conventions.

In fact, the invention of a convenient and flexible notation is often more seminal than is a proof of even the deepest of theorems. A kind of coarse-grained history of mathematics could even be written which would be devoted entirely to the introduction of significant new notations. Roman numerals (and their clumsy Greek cousins), despite their lingering and pretentious use in contexts like Super Bowl XXIV and Copyright MCMLXXXIX, do not lend themselves to calculation as do the Arabic numerals which replaced them. (See the entry on *Arabic numerals.*) Discussion of individual numbers and magnitudes cannot be easily generalized without the introduction of a notation for variables (invented by François Viète). Likewise, it is difficult to symbolically manipulate geometric figures lacking the notational tools of analytic geometry (René Descartes and Pierre Fermat).

There were really no formal theorems connected with the introduction of the above symbolisms, but they most certainly were not merely cosmetic. Their introduction and subsequent universal adoption codified the insights and ideas of some very smart and learned folks and made them accessible to all. In a few minutes' time, fourth-graders (or should I say IVth-graders?) can perform computations it used to take medieval European professors hours to complete. Utilizing variables, ninth-graders can set up and solve algebraic equations whose solutions in some cases were unknown to both ancient Greek mathematicians and these same medieval professors. A college freshman using analytic geometric techniques can develop insights and establish properties of geometric figures which eluded those restricted to purely classical means of expression.

In a similar vein the notation of set theory, prevalent in every area of mathematics, is simple enough to teach in elementary schools. Despite this, the preface or introductory chapter of innumerable texts at every level includes ten to twelve pages whose content is roughly that of the next few sentences. We write $p \in F$ to indicate that p is a member of the set F and $F \subset G$ to mean that every element of F is also an element of G. Given two sets A and B, $A \cap B$ is the set containing those elements that belong to both A and B; $A \cup B$ the set containing those elements belonging to A, to B, or to both; and A' the set of those elements not belonging to A. The empty set, a set containing no elements, is indicated by $\emptyset$ or sometimes by $\{\}$—braces enclosing nothing. End of mini-course.

The above notations as well as those which have since been developed—those for derivatives, integrals, and series in calculus, notations for operators and matrices in a variety of mathematical disciplines, symbols for connectives and quantifiers in logic, the field of category theory, which is in a sense nothing but notation—all possess three properties (A, 2, and III) which explain their adoption and use. They are suggestive, the relation between the symbols corresponding in some natural way to the relation between mathematical objects (Y^5, for example, is $Y \times Y \times Y \times Y \times Y$). Generally, too, good symbolisms are manipulable and utilize some associated rules or algorithms (whether they be for multiplying numbers or solving linear systems of differential equations). Finally, effective notations are encapsulating, containing much critical information in a compact form without extraneous or redundant elements (such as the previous prepositional phrase).

Of course, notation as a formalized system of representation transcends mathematics. Double-entry bookkeeping (in which every transaction is entered as both a debit and a credit) transformed the practice of accounting, chemical symbols greatly simplified the recording of chemical reactions, and Feynman diagrams of subatomic interactions clarified quantum mechanics. There are myriad other examples of notational systems, the most fundamental, pervasive, and all-encompassing one being the alphabet and the written language it made possible.

OULIPO—MATHEMATICS IN LITERATURE

The Ouvroir de Littérature Potentielle (Workshop of Potential Literature), Oulipo for short, is the name of a small group of primarily French writers, mathematicians, and academics devoted to the exploration of mathematical and quasi-mathematical techniques in literature. Founded by Raymond Queneau and François Le Lionnais in Paris in 1960, the group searches for new literary structures via the imposition of unusual constraints, for methods of systematically transforming texts, and for ways to exemplify mathematical concepts in words. The group has put out various manifestos, but as with yoga enthusiasts, their techniques and results are more interesting than their philosophy.

Queneau's *100 Trillion Sonnets* is a prime example of Oulipo's combinatorial approach to literature. The work consists of ten sonnets, one on each of ten pages. The pages are cut so as to allow each of the fourteen lines of each sonnet to be turned separately. Thus any of the ten first lines may be combined with any of the ten second lines, resulting in 10^2 or 100 different pairs of opening lines. Any of these 10^2 possibilities may be combined with any of the ten third lines to yield 10^3 or 1,000 possible sets of three lines. Iterating this procedure, we conclude that there are 10^{14} possible sonnets. (See the entry on *the multiplication principle.*) Queneau claimed that they all made sense, although it's safe to say that the claim will never be verified, since there are vastly more texts in these 10^{14} different sonnets than in all the rest of the world's literature.

Another good example of Oulipo's work is Jean Lescure's (N + 7) algorithm for transforming a text. Take an excerpt from your favorite newspaper, novel, or holy book and replace each noun in it with the seventh noun following it in some standard dictionary. If the original is well written, the resulting text usually retains its rhythm and occasionally even something of its sense. "In the behavior goddaughter created the heavyweight and the earthquake. And the earthquake was without format, and void; and dark horse was upon the facia . . ." Of course, we may modify the algorithm by replacing every other noun or by taking the tenth noun after the word in the dictionary, etc.

Yet another quintessentially Oulipian work, Georges Perec's 300-page novel *La Disparition,* doesn't contain a single letter "e" except, of course, for the four unfortunate instances in his name. Think of this: no "the," "are," "were," "he," "she," "they," nor even an "even." In an essay on such lipograms, works that omit letters, Perec defends the sanity and seriousness of such an undertaking, arguing that constraint and artifice are the engines that have driven not only Oulipians but also many mainstream authors (among them François Rabelais, Laurence Sterne, Lewis Carroll, James Joyce, Jorge Luis Borges, and Oulipo member Italo Calvino) to plumb all of a language's possibilities.

Some of these possibilities derive from ways to merge texts by "multiplying" one text by another (like matrices of numbers), from finding the logical intersection of two disparate works, or from the "haikuifying" of a long poem to make a shorter one. More familiar bits of wordplay are not ignored either. Palindromes, written locutions that read the same backward and forward (the classic example: A man, a plan, a canal, Panama); spoonerisms, transpositions of sounds in two or more words (Time wounds all heels, Tee many martoonis, or the junior high school favorite Chuck you, Farlie); the American Oulipo member Harry Mathews's perverbs, the combining of two proverbs into one (A rolling stone gets the worm, or A bird in the hand waits for no man); snowball sentences, each of whose words is one letter longer than its predecessor (I do not pass Sally unless feeling reckless); and all of their near and distant cousins pervade Oulipo's poems, stories, and novels.

Strangely enough, until recently Oulipo has not shown as much interest in computers as their penchant for syntactical permutation might suggest. Adapting various word-processing programs with special dictionaries, thesauruses, key word counters, and what have come to be called hypermedia tools would make their combinatorial literary play easier. (See the entry on *human consciousness, its fractal nature.*)

When they're good, Oulipo's works are fresh, stimulating, and fun. When they're not, the incessant punning and trickery is as tiresome as babysitting for a brilliantly loquacious eight-year-old. (Note that I've accomplished the awesome feat of writing this entire entry with only twenty-five letters—not a single instance of the last letter of the alphabet.)

PARTIAL ORDERINGS AND COMPARISONS

∞

The mathematical structure known as a partial ordering better describes most humanly interesting phenomena than do other orderings. A partial ordering is any set with an ordering (i.e., some elements of the set are greater than others) that allows for some pairs of elements to be incomparable. It is to be contrasted with a linear or total ordering such as "is at least as tall as" or "is at least as heavy as." In the latter cases, it's clear that of any two people one is taller than the other, one heavier. In partial orderings two elements may simply be incomparable with respect to the given ordering relation.

To illustrate, consider the set of circles lying in a plane. Each of these circles contains and is contained in other circles, but given any two circles selected at random, it's likely that neither contains or is contained in the other. Most pairs of circles are incomparable, and the relation "contains" is a partial ordering. Properties such as beauty, intelligence, even wealth, are less simplistically discussed in terms of partial orderings than in terms of linear or total orderings.

In fact, trying to convert a partial ordering into a total one is, I think, at the root of many problems. Reducing intelligence to a linear ordering—a number on an IQ scale—does violence to the complexity and incomparabilities of people's gifts. Likewise with a beauty or wealth index. Trying to prioritize our preferences in a large field of political candidates leads to various voting paradoxes (see the entry on *voting systems*), and the term "political spectrum" is simplistic and reductionistic. Similar difficulties arise if we rank our friends. Most personal and public issues that face us are sufficiently complicated and multidimensional that an attempt to force them into a procrustean list is myopic and narrow-minded (the latter literally so).

Nevertheless, lists are appealingly simple. People always want to know who's the "top man" in this or that medical specialty (the connection between sexism and linear hierarchies is not accidental), who's making the most money, who's number one on the best-seller list. I wonder sometimes if those with a more balanced, harmonious approach to life aren't at a disadvantage in a world where obsession and monomania seem to pay off so handsomely. Maybe a controlled or limited obsession, although this sounds like an oxymoron, is the appropriate response.

(There is a lot of talk about the differences in mathematics performance between males and females, in particular about the numbers of each who pursue higher mathematics. There is no convincing evidence for a genetic basis to these differences. They are due, I suspect, to socialization and *perhaps* to genetically influenced personality differences. In this regard, one possibly relevant observation: Although I know a number of absolutely first-rate women programmers, I've met very few women hackers of the sort that regularly stay up all night writing pointless computer code, have dirty hair and clothes and bags under their monitor-glazed eyes, have no friends, subsist on potato chips, candy bars, and caffeine drinks, change their system configurations hourly, and generally disappear into their self-created electronic empires. Women programmers, and I realize this may sound patronizing, are generally too balanced to be computer hackers. It may be that mathematical research, although it doesn't require quite the same compulsiveness and monomaniacal immaturity as does hacking, is more likely to appeal to those with personality characteristics traditionally deemed masculine.)

Returning to partial orderings, I maintain that when we must compare things, a tree or a bush is often a better model than a pole. Trees and bushes allow for incomparable elements (on different branches) as well as for comparable ones (along a single branch), while poles collapse everything into one dimension.

PASCAL'S TRIANGLE

Pascal's triangle is a triangular array of numbers which was known to the Chinese in 1303, more than 300 years before the French mathematician and author Blaise Pascal discovered many of its most interesting properties. Each number in the triangle except for the 1's on the edges is obtained by adding the two numbers above it. Thus if we were to add another row to the figure below, it would be 1, 7, 21, 35, 35, 21, 7, 1.

Although the above rule is simple enough, the diversity of patterns contained in the triangle is surprising. Probably the most important of these patterns involves the combinatorial (or binomial) coefficients. These numbers tell us, among other things, how many poker hands, lottery tickets, and Baskin-Robbins triple-scoop ice-cream cones there are. In general, they give us the number of different ways of choosing R elements out of N.

Let me illustrate with a small number. Consider the fourth row of the triangle—1, 4, 6, 4, 1. These five numbers indicate the number of ways of choosing 0, 1, 2, 3, and 4 elements, respectively, out of a set of 4 elements. Thus, the first number in the row, 1, indicates the number of ways of selecting 0 elements out of a set of 4. There's only one way to do this: Don't take any of them. The next number, 4, indicates the number of ways of selecting 1 of the 4 elements. We observe astutely that there are four possibilities here: Choose the first element, the second, the third, or the fourth. The third number in the row, 6, indicates the number of ways there are of selecting 2 out of the 4 elements. To illustrate this, assume that the 4 elements are the letters A, B, C, D. Then the six ways of selecting exactly 2 of them are: AB, AC, AD, BC, BD, and CD. The next number in the row, 4, is the number of ways of choosing 3 out of the 4 elements. There are four ways to do this: Simply decide which one of the 4 not to select. The last number is 1, since there is only one way to select 4 elements out of a set of 4: Just take them all.

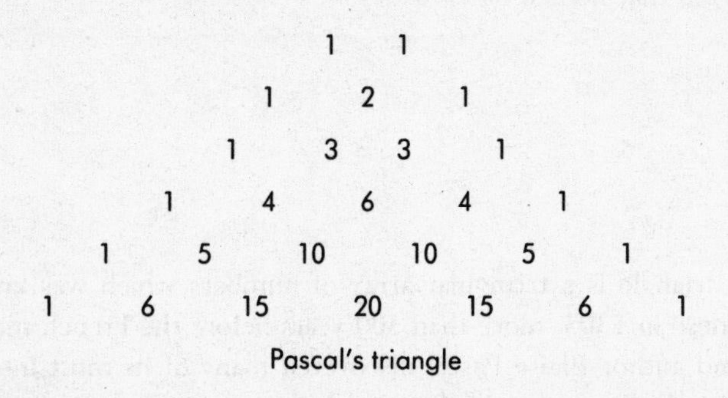

Pascal's triangle

All rows work the same way. The numbers in the sixth row—1, 6, 15, 20, 15, 6, 1—represent, respectively, the number of ways of selecting 0, 1, 2, 3, 4, 5, and 6 elements out of a set of 6 elements. We see that there are 15 ways of choosing 2 elements out of 6, and 20 ways of choosing 3. If we examine the 31st row in Pascal's triangle and count over 4 places, we find the number of possible ways to choose 3 flavors from among Baskin-Robbins' 31 flavors: 4,495. If we examine the 48th row and count over 7 places, we find the number of ways of choosing 6 elements out of 48—i.e., the number of possible lottery tickets in a pick-six lottery: 12,271,512. If we examine the 52nd row and count over 6 places, we find the number of ways of choosing 5 elements out of 52—i.e., the number of possible poker hands: 2,598,960.

(I feel tempted to give more examples, but whenever I go on like this, citing example after similar example, the memory of my freshman calculus professor induces me to control my pedantic compulsions and stop. This professor taught all 200 of us in a large lecture hall. One day he was behaving a little erratically, but this wasn't too unusual for him, so we ignored it. He was talking about conditionally convergent series and wrote on the left side of the board something like 1 − 1/2 + 1/3 − 1/4 + 1/5 − . . . . He proceeded in this way and when he reached the middle of the very wide blackboard he was up to 1/57 − 1/58 + ... The class thought this was another example of his eccentricity and seemed amused, but when he neared the right side of the blackboard and continued . . . − 1/124 + 1/125 − . . . , we grew quiet. Finally he reached the end of the blackboard and turned around and faced us. His hand holding the chalk quivered for a bit and then he dropped the chalk and left the stage. He never returned to class, and it was reported that he had had a nervous breakdown.)

This way of generating the so-called combinatorial coefficients is not very practical for large numbers, so a formula derived from the multiplication principle (see the entry on *the multiplication principle*) is used. It says that the number of ways of choosing R elements out of N, usually symbolized C(N,R), is N!/R!(N − R)!, where, for any number X, X! indicates the product of X, X − 1, X − 2, and so on, down to 1. For example, 6! = 6 × 5 × 4 × 3 × 2 × 1 = 720. Checking that the formula gives the same answer as Pascal's triangle does when N = 6 and R = 2, we note that 6!/2!4! = 15, the number of ways of choosing 2 elements from a set of 6; C(6,2) = 15.

This formula is particularly valuable in probability theory and combinatorics, where counting arguments are common. For example, if you are completely ignorant of a subject but must take a multiple-choice test having 12 questions each with 5 possible answers, the probability of your answering the first 4 correctly and the next 8 incorrectly is $(1/5)^4 \times (4/5)^8$. This is also the probability you will answer questions 3, 4, 7, and 11 correctly and the rest incorrectly, or questions, 1, 6, 7, and 9 correctly and the rest incorrectly. To find the probability of answering any set of exactly 4 questions correctly, we find the number of different ways of choosing 4 out of 12—this is C(12,4) or 495—and multiply this number by $(1/5)^4 \times (4/5)^8$, the probability of answering any particular set of 4 questions correctly. The result, $C(12,4) \times (1/5)^4 \times (4/5)^8$, is the answer (about 13%) and also a special case of the important binomial distribution in probability theory.

The numbers in the Nth row of Pascal's triangle, it should also be mentioned, are the coefficients in the expansion of $(X + Y)^N$. To illustrate, recall (or take my word for it) that $(X + Y)^2$ is equal to $\underline{1}X^2 + \underline{2}XY + \underline{1}Y^2$ and that $(X + Y)^3$ equals $\underline{1}X^3$ +

$\underline{3}X^2Y + \underline{3}XY^2 + \underline{1}Y^3$, and note that the underlined coefficients match the numbers in the 2nd and 3rd rows of Pascal's triangle. And $(X + Y)^4 = \underline{1}X^4 + \underline{4}X^3Y + \underline{6}X^2Y^2 + \underline{4}XY^3 + \underline{1}Y^4$.

A few of the other patterns lurking within Pascal's triangle are: the whole numbers along the penultimate diagonals, the triangular numbers (expressible as triangular arrays of dots—3, 6, 10, and so on) along the next diagonals toward the center, the tetrahedral numbers (expressible as tetrahedral arrays of dots—4, 10, 20, and so on) along the next diagonals, higher-dimensional analogues of these along still more interior diagonals, and the Fibonacci numbers as sums of the diagonal elements which proceed upward (downward) from one row to the preceding (succeeding) one. Locating these and other configurations gives one an appreciation for the beauty and complexity that is sometimes inherent in the simplest of rules.

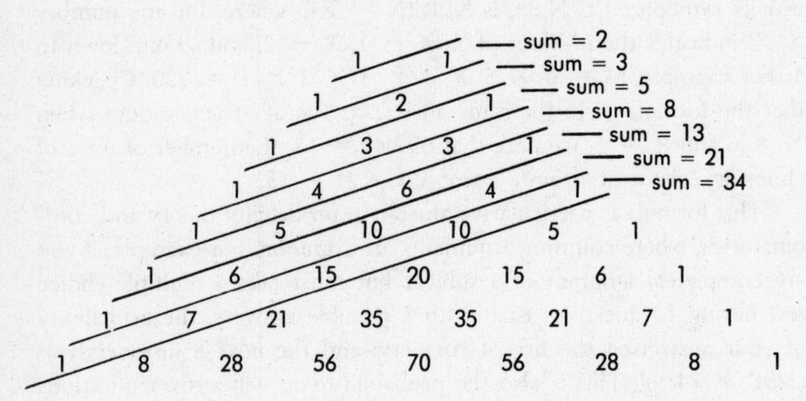

The Fibonacci numbers arise from Pascal's triangle.

PHILOSOPHY OF MATHEMATICS

∞

What are numbers, points, probabilities? What is the nature of mathematical truth? Why is mathematics useful? These are some of the questions whose answers you won't find here. I will try, however, to give a taste for a couple of the issues involved.

The most casual observer notes that mathematical theorems are not confirmed in the same way that physical laws are. They seem to be necessary truths, while statements from the empirical sciences (physics, psychology, and cooking) seem quite contingent on the way the world actually is. At least conceptually, Boyle's gas laws and the history of the Austro-Hungarian Empire could easily have been otherwise; not so the proposition that $2^5 = 32$.

But where does the certainty and necessity of mathematical truth come from? Working mathematicians don't usually concern themselves with this question, but, if pressed, most of them would probably say something like the following: Mathematical objects exist independently of us, and statements about them are true or not independently of our knowledge of them or our ability to prove them. Presumably these objects exist in some Platonic realm beyond time or space. If this is the case, however, how do we find out about such objects and the facts regarding them?

Immanuel Kant's answer was that mathematics (or at least its basic axioms) was knowable a priori by intuition alone and that its necessity was self-evident. Contemporary intuitionists, while not subscribing to

Kant's ideas about space, time, and number, also ground the necessity of mathematics in the indubitability of simple mental activities. Some even discredit proofs of an object's existence unless there is a constructive process by which it might be found.

Most other philosophers of mathematics are disturbed both by the subjectivism of Kant and by the untenability of naïve Platonism. The so-called logicists, Bertrand Russell, Alfred North Whitehead, and Gottlob Frege, tried to show that mathematics could be reduced to logic and thus was as certain as the simple proposition "A or not A," that in fact mathematical statements were just circuitous ways of saying "A or not A." They hoped in this way to find a warrant for the certainty of mathematical statements, but they weren't completely successful. What they called logic contained ideas from set theory that were just as problematic as the mathematical statements which followed from them.

The conventionalist response to the question was that mathematics attained its necessity by convention, fiat, and definition. Its truths were just matters of convention and thus weren't any more obscure than the fact that 3 feet equals 1 yard. In a related approach, formalist philosophers maintained that mathematical statements are not about anything, but rather should be considered meaningless sequences of marks which are governed by rules, much as the movement of pieces on a chessboard is governed by the rules of chess. That a knight moves two squares in one direction and one in a perpendicular direction is necessary but not mysterious.

The smug dismissiveness of these latter positions is initially attractive, but they fail utterly in accounting for what Nobel laureate Eugene Wigner called "the unreasonable effectiveness of mathematics" in describing reality. Other philosophers counter that the correspondence between mathematical structures and physical reality is not "unreasonable" at all. It is, they claim, not radically different from the reasonable correspondence between various biological senses (olfactory, taste, auditory, tactile, and visual) and aspects of physical reality, mathematical perception being a sort of abstract sixth sense.

What the source of mathematical necessity is and whether numbers are mental constructs, facets of an idealized reality, or just rule-governed markings are issues which, in various guises, have resonated all through the history of philosophy. In the Middle Ages, for example, the battle was among the idealists, realists, and nominalists, whose positions on the nature of universals such as Redness or Triangularity were somewhat analogous to those taken by the intuitionists, logicists (or Platonists), and formalists today. (See also the entries on *non-Euclidean geometry, probability,* and *substitutability.*)

These concerns transcend mathematics. They're intimately linked, for example, to the philosophical distinction between analytic and synthetic truths. An analytic statement is true by virtue of the meanings of the words it contains, whereas a synthetic statement is true or not by virtue of the way things are. (A special sort of analytic statement, a logically valid statement, is true by virtue of the meanings of the logical words "and," "or," "not," "if . . . , then . . . ," "some," and "all." Statements true by virtue of the first four of these alone are termed tautologies. See the entries on *quantifiers* and *tautologies.*)

Thus, "If Waldo is smelly and toothless, then he's smelly" is analytic, whereas "If Waldo is smelly, then he's toothless" is synthetic. The same opposition is exemplified by "Bachelors are unmarried men" vs. "Bachelors are lecherous men" and by "UFOs are flying objects that haven't been identified" vs. "UFOs contain little green creatures." Philosophers count mathematical truths as analytic and most others as synthetic, and, though not immutable, the distinction is a handy conceptual device. When Molière's pompous doctor announces that the sleeping potion is effective because of its dormitive value, he is making an empty, analytic statement, not a factual, synthetic one.

Revolving as it does about issues of transcendent truth and certainty, the philosophy of mathematics also has a certain natural resonance with religious thought. Why else would I, a devout agnostic, have used words like "divine," "priestly," "heaven," "purity," and "reverence" so frequently in these entries? The similarities are largely metaphorical, but sometimes metaphors determine attitudes and actions.

Finally, whatever "ism" one subscribes to (or hides from), Gödel's incompleteness theorems (see the entry on *Gödel*) shuffled the philosophical cards and caused all parties to recompute their hands. The existence of unprovable statements, for example, indicates that the truth of such statements cannot reside solely in their proofs from axioms. Moreover, the very consistency of a mathematical theory is one of the statements which can't be proved but must simply be assumed (or taken on faith, if you like that language).

PI

∞

I once took an informal survey of nonmathematician friends and neighbors of mine to determine how many knew what pi was. Almost all knew it had something to do with geometry and with circles in particular, the latter fact undoubtedly explaining why some thought it was spelled p-i-e. (The Greek letter π, pi in English, has been in use to denote the quantity since the eighteenth century.) A minority knew its value was about 22/7 (close enough for full credit), but most estimates were way off (including an eminent lawyer's sonorously delivered one of 5.42). Some said π was the area of a circle, while the majority tried to conceal their ignorance and/or my impertinence with a joke (the lawyer asked me to recite the provisions of a bankruptcy statute). Only a few knew it to be the ratio of a circle's circumference to its diameter. That is, π equals C/D, the circumference divided by the diameter.

Ancient estimates for π include 3 (Old Testament—an insurmountable problem, it would seem, for biblical literalists), 25/8 (Babylonian), 256/81 (Egyptian), 22/7 (Greek), 355/113 (Chinese—correct to six decimal places), and $\sqrt{10}$ (Indian—a pleasing coincidence). Considerably more precise is 3.14159265358979323846264338327950-28841972, but π is an irrational number (not expressible as a ratio of two whole numbers), which implies that its decimal expansion goes on forever without repeating itself. It is a transcendental number, which means that it is not the solution of any algebraic equation. As one of the most fundamental constants in mathematics, π figures in many important formulas, a very basic one being that the area of a circle equals π times the square of the circle's radius ($A = \pi R^2$). The volume of a sphere, by constrast, equals 4/3 times π times the cube of the sphere's radius ($V = 4/3 \times \pi R^3$). Pi is indispensable to the statement of Scottish physicist James Clerk Maxwell's celebrated laws of electromagnetism and also appears in many formulas and contexts where its appearance is more surprising, there being no circles or spheres about to which to attribute its occurrence. For example, $\pi/4 = 1 - 1/3 + 1/5 - 1/7 + 1/9 - 1/11 + \ldots$, and $\pi^2/6 = 1/1^2 + 1/2^2 + 1/3^2 + 1/4^2 + 1/5^2 + \ldots$.

The Buffon needle problem first posed in the eighteenth century by the Comte de Buffon does not involve circles either, but its solution nevertheless contains π and was in fact once used to calculate π. Suppose with Buffon that we have a floor made up of parallel strips of wood each 3 inches wide. Assume further that we have a needle 3 inches long and that we toss the needle carelessly onto this striped floor. What is the probability that the needle will land so as to cross a line between two adjacent strips? Equivalently, what are the chances that the needle will not be contained entirely on a single strip of wood? I will skip the derivation, but the probability can be shown to be $2/\pi$.

One could now use this result to estimate π in the following way. (See the entry on *Monte Carlo method of simulation.*) Toss the needle onto the floor 10,000 times (imagine yourself devoted to experimental mathematics or else in prison with nothing to do) and determine what fraction of the time it crosses a line. Say it crosses 6,366 times and that our estimate of the probability of the needle's crossing is therefore .6366. (The needle will cross a line approximately this often provided it is as likely to land in one spot on the floor as another and that its orientation is as likely to be one angle as another.) If we set this latter probability equal to $2/\pi$ and solve the resulting equation for π, we come up with 3.1417 as our estimate for π in this case, a number quite close to the real value. There are needleless to say incomparably better ways to ascertain the value of π.

The appeal of π is due, I think, to its universality. Like Mount Everest, it's always there challenging us. The latest (as of 1990) supercomputer calculations of π contain more than a billion digits. Expressed in a less anthropocentric binary (base 2) number system, π might even be used, as many science fiction stories have suggested, as a way to communicate our technological sophistication and friendly Euclidean nature to distant beings. Even if the more innumerate among them interpret the signal as some sort of interstellar box score, the message will get through.

Finally, a little πuzzle: Imagine a string tied tautly around the earth at the equator. What length of extra string would have to be added to this string so that the extended string might lie one foot above the earth's surface all the way around the equator?

The initially surprising answer is that a little more than six feet would suffice. The explanation depends upon the formula for the circumference of a circle: $C = \pi D$. If the diameter of the earth is 8,000 miles, then a length of string equal to $\pi \times$ 8,000 miles, or approximately 25,120 miles, is needed to circle the equator. If we stipulate that the string must lie one foot above the earth's surface all the way around the equator, we're asking that the diameter be increased by two feet. Thus the larger circumference is $\pi \times$ (8,000 miles + 2 feet), which equals ($\pi \times$ 8,000 miles) + ($\pi \times$ 2 feet), or approximately 25,120 miles plus 6.28 feet. Hence only 6.28 feet of extra string is required.

PLATONIC SOLIDS

I once had a student in an introductory survey course who on the final exam referred to Platonic solids as Caesar salads. She was otherwise rather nondescript, and I was never certain whether her remark was intentional humor or a manifestation of "math anxiety." In any case, Platonic solids are three-dimensional solids whose polygonal surfaces are all congruent and whose corners all meet at the same angles. A cube is a Platonic solid because its surfaces are all squares of the same size, whereas a shoebox isn't because its sides are not all congruent. Another example of a Platonic (or regular) solid is the tetrahedron whose four sides are equilateral triangles of the same size. (Milk cartons in some countries are tetrahedral.) I can't give you too many more examples because, as the early Greek geometers discovered, there are only five Platonic solids.

The five are the aforementioned cube and tetrahedron, the octahedron whose eight sides are equilateral triangles of the same size, the dodecahedron whose twelve sides are regular (equilateral and equiangular) pentagons, and the icosahedron whose twenty sides are all equilateral triangles of the same size. The beauty and paucity of these regular solids (why just these five?) prompted a mystical reverence for them which resulted in the astronomer Johannes Kepler's attempts to use properties of the solids to explain the motion of the planets in our solar system. (Fortunately he stayed at the job and later came up with planetary laws based on observation rather than a priori mathematics.) Even today some people meditate within large tetrahedral structures or caress and fondle symmetric crystals in the belief that understanding or health or some such desideratum will somehow result. Interestingly, some of the discussion surrounding the shapes of the seven basic discontinuities studied in modern catastrophe theory has this same overwrought, pseudoscientific character.

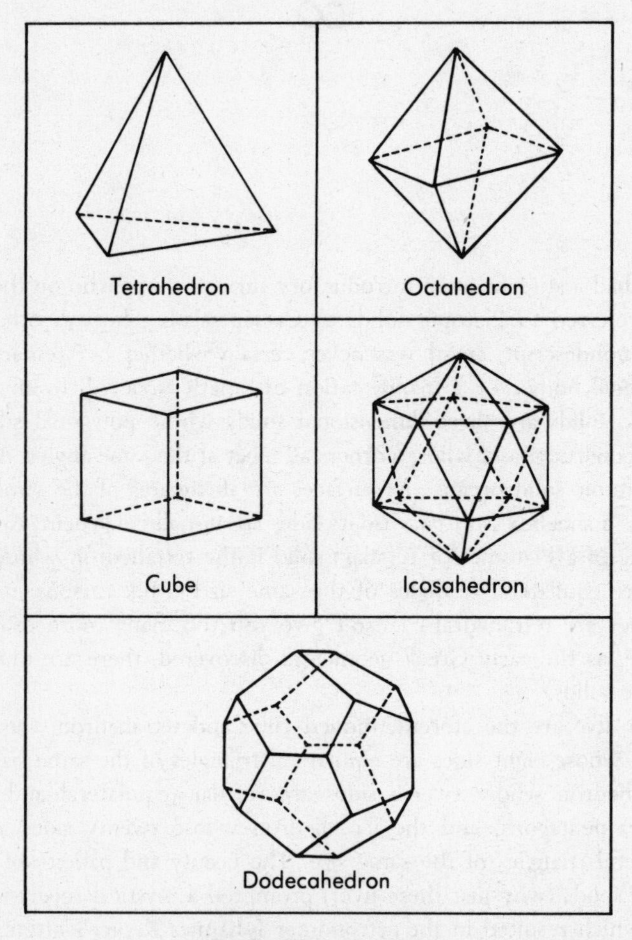

The five Platonic solids

Mysticism aside, there is something pristine about the Platonic solids. That there are only five can be proved easily from a well-known formula due to the eighteenth-century Swiss mathematician Leonhard Euler. The formula states that for any polyhedron (a solid whose surfaces are polygons not necessarily of the same size or shape) the result of adding the number of vertices to the number of faces and subtracting from this sum the number of the polyhedron's edges is always 2. More mathematically phrased and making use of the obvious abbreviations, the relation is $V + F - E = 2$. (It's helpful to verify that the formula is true for cubes and shoeboxes, in both of which cases $V = 8$, $E = 12$, and $F = 6$.) Since the formula holds for all polyhedrons (Platonic or otherwise) and since a solid's being Platonic imposes additional constraints on V, E, and F, a little computation demonstrates that only five Platonic solids are possible.

Surprisingly, formulas for the surface area and volume of Platonic solids involve the number e, the base for natural logarithms. There is also an unexpected connection between the golden rectangle (see the entries for *e* and the *golden rectangle*) and the icosahedron: If each of three equal golden rectangles intersects the other two symmetrically and perpendicularly, the corners of these rectangles are the vertices of a regular icosahedron. A more comprehensible fact about Platonic solids is that they make good randomizing devices—cubes (i.e., dice) for choosing any of six outcomes with equal probability, dodecahedrons any of twelve, and so on. In fact, the elemental nature of the Platonic solids (like that of pi and e) makes them good candidates for communicating with other intelligent beings should there happen to be any.

Platonic solids are one of the few interesting topics in classical solid geometry, a once moribund subject whose higher-dimensional generalizations lie at the forefront of research in several areas of physics, abstract algebra, and topology.

PRIME NUMBERS

∞

Before the advances in nuclear physics which revealed the atom to be a Balkanized society of subatomic parts, prime numbers were often compared metaphorically to atoms. Being made of sturdier stuff (or rather of sturdier nonstuff) than physical atoms, prime numbers retain their timeless indivisibility. They are to be contrasted with composite numbers, which can be expressed as a product of two smaller numbers. Prime numbers cannot be so expressed. The numbers 8, 54, and 323 are composite since they equal, respectively, 2×4, 6×9, and 17×19, while the numbers 7, 23, and 151 are prime since they can't be further broken down or factored. The first dozen prime numbers are 2 (the only even prime—why?), 3, 5, 7, 11, 13, 17, 19, 21 (just kidding), 23, 29, 31, 37. It can be shown that every whole number can be factored into prime numbers in exactly one way. The number 60, for example, is equal to $2 \times 2 \times 3 \times 5$; 1,421,625 equals $3 \times 5 \times 5 \times 5 \times 17 \times 223$; and 101, being prime, simply equals itself.

Because of their simplicity and almost tangible thereness, prime numbers have fascinated men since before the time of the ancient Greeks. One question that occurs quite naturally is: How many prime numbers are there? If you were to continue the list of primes begun above, you would note that the primes becomes sparser and sparser the further out one explores. There are more primes between 1 and 100 than there are between 101 and 200. Thus one might guess that there is a largest prime just as there is an element having a largest atomic number. Euclid showed, however, that there is no largest prime and that therefore there are an infinite number of primes.

Euclid's demonstration of this is such a beautiful example of what is often called an indirect proof that I will risk arousing your mathematical anxiety and reproduce it. We assume at the outset that there are only finitely many primes and try to derive a contradiction from this assumption. Thus, we list the prime numbers 2, 3, 5, . . . , 151, . . . P; P we will take to be the largest prime number. Now form a new number N by multiplying all the primes in the above list together. Thus $N = 2 \times 3 \times 5 \times \ldots \times 151 \times \ldots \times P$.

Consider the number (N + 1) and whether 2 divides it evenly (with no remainder) or not. We see that 2 divides N evenly since it is a factor of N. Therefore 2 does not divide (N + 1) evenly, but leaves a remainder of 1. We see also that 3 divides N evenly since it too is a factor of N. Therefore 3 does not divide (N + 1) evenly, but also leaves a remainder of 1. Similarly for 5, 7, and all the prime numbers up to P. They each divide N evenly, and therefore each leave a remainder of 1 when divided into (N + 1).

What does this mean? Since none of the prime numbers 2, 3, 5, . . . , P divides (N + 1) evenly, the number (N + 1) is either itself a prime number larger than P, or it is divisible by some prime number that is larger than P. Since we assumed that P was the largest prime number, we have a contradiction: We have established the existence of a prime number larger than the largest prime number. Therefore our original assumption that there are only finitely many prime numbers must be false. End of proof. QED.

A more difficult proposition, the prime number theorem, tells us roughly how frequently primes turn up throughout the integers. If P(N) is the number of primes less than or equal to N [note that P(10) equals 4 since there are four primes—2, 3, 5, and 7—less than or equal to 10], then the theorem states that as N increases, the ratio N/P(N) gets closer and closer to the natural logarithm of N. Using this result, for example, we can calculate that approximately 3.6% of the first trillion numbers are prime. There is, I think, a kind of immutable purity about these and other theorems and proofs in number theory that seems, at least when one is in a sufficiently pensive mood, almost divine.

Another quasi-divine aspect of prime numbers is the ease with which one can make simple statements about them whose truth or falsity is unknown. One example is the Goldbach conjecture, which asserts that every even number greater than 2 is the sum of two primes. We check that $4 = 2 + 2$, $6 = 3 + 3$, $8 = 3 + 5$, $10 = 5 + 5$, $12 = 5 + 7$, and so on up to $374 = 151 + 223$ and beyond. The conjecture is believed true, but has not been proved. Also unknown is whether there are an infinite number of twin primes, primes such as 17 and 19 or 41 and 43 or 59 and 61 which differ by 2. Again the proposition is believed to be true but it has yet to be proved.

Despite the divinity metaphor, these pure and seemingly useless musings about primes have relevance to credit cards, telecommunications, and national security. The basic idea is that it's elementary to find the product of two 100-digit primes, but almost impossible to factor the resulting 200-digit number. We can utilize this and other properties of primes and the latter monstrously large numbers to encode messages that can only be decoded by someone who knows their prime factors. Such codes are used daily by banks to transfer funds, and versions of them for military and intelligence purposes are produced at the top secret National Security Agency. Such applications appear as incongruous as a monk working in a munitions factory. Holy Euclid!

P.S. As of 1990, the number $[(391{,}581 \times 2^{216{,}091}) - 1]$ is the largest known prime. The search required a supercomputer working on and off for more than a year.

PROBABILITY

∞

Everyone has an intuitive grasp of probability. Sometimes it's as primitive as that of the barber who once revealed his lottery strategy to me: "The way I figure it, I can either win or lose, so I've got a 50-50 shot at it." Despite many blind spots (see the entry on *coincidences*), however, people's visceral assessments of probability are usually considerably more sophisticated. Phrases like "the probability of the coin's landing heads," "the likelihood Martha will marry George," and "the chances it will rain for the game tomorrow" trip easily off our tongue, and most seem to know what they mean. It's only when we ask what probability actually is that we feel totally baffled and at a loss.

The question is not easy to answer, but there have been a number of attempts at it. Some people have conceived of probability as a logical relation, as if one could just glance at a die, note its symmetry, and decide by logic alone that the probability of a 5 turning up must be 1/6. Others have suggested that probability is merely a matter of subjective belief, nothing more than an expression of personal opinion. Relative frequency, according to others, is the key to the analysis, the probability of an event being a shorthand way of indicating the long-run percentage of times it occurs, "long-run" usually remaining unexplained.

There are still other versions and other views, but none is universally cogent. What's finally happened is that mathematicians have simultaneously retreated and declared victory. They've observed that since, by all reasonable definitions, probability ends up possessing certain formal properties, probability should simply be defined to be whatever satisfies these formal properties. Not very philosophically gratifying, but mathematically liberating.

Note that this is similar to what occurred in geometry with points and lines. Euclid gave vacuous little definitions of these notions which he never really used, whereas more modern approaches to plane geometry have been axiomatic, defining points and lines to be whatever satisfies the properties set forth in the axioms. The Russian mathematician A. N. Kolmogorov is the originator of this abstract approach to probability, but rather than describe his elegantly bare formulation, let me list a few of the most basic properties and theorems of probability, most of them known since the gambling origins of probability theory in the seventeenth century.

First, probability is a number between 0 and 1, 0 indicating impossibility; 1, certainty; and intermediate numbers, intermediate degrees of likelihood. Equivalently, we may take the range to be between 0% and 100%. If two or more events are mutually exclusive, the probability that some one of them will occur is obtained by adding the individual probabilities. Thus, the probability a randomly selected earthling is Chinese, Indian, or American is approximately 45% (25% Chinese added to 15% Indian added to 5% American).

For two arbitrary events (similar formulas hold for three or more events), the probability that at least one of them occurs is a little more difficult to obtain: Add the probabilities of the two events and then subtract from this sum the probability that both events will occur. If in a certain large apartment building in New York 62% of the tenants read *The New York Review of Books,* 24% read the *National Enquirer,* and 7% read both, and one picks a tenant at random, then the probability that he or she reads at least one of the two periodicals is 79% (62% + 24% − 7%). The probability an event won't occur is 100% minus the probability it will occur. Hence a 79% chance of reading at least one of the periodicals indicates a 21% chance of reading neither.

Of critical importance in probability theory is the notion of independence. Two events are said to be independent when the occurrence of one of them does not make the occurrence of the other more or less probable. If one flips a coin twice, each flip is independent of the other.

If one rolls a pair of dice, the top face of one die is independent of the top face of the other. If one chooses two people from the phone book, the height of one is independent of that of the other.

Calculating the probability of two independent events occurring is quite easy to do—you simply multiply their respective probabilities. Thus, the probability of obtaining two heads is $1/4 - 1/2 \times 1/2$. The probability of rolling a 2 with a pair of dice, i.e., (1,1), is $1/36 - 1/6 \times 1/6$, while the probability of rolling a 7 is 6/36 since there are six mutually exclusive ways [(1,6), (2,5), (3,4), (4,3), (5,2), (6,1)] in which the numbers on the faces of the dice can add up to 7, and each of these ways has probability $1/36 - 1/6 \times 1/6$. The probability of two people chosen from the phone book both being over 6 feet tall is obtained by squaring the probability that a single person so chosen is over 6 feet tall.

This multiplication principle for probabilities may be extended to sequences of events. The probability of a die's turning up 3 on four consecutive rolls is $(1/6)^4$; of a coin's landing heads six times in a row, $(1/2)^6$; of someone's surviving three shots in Russian roulette, $(5/6)^3$. If a book is such that only 10% of its readers will judge it positively (90% finding it loathsome), then the probability that each of a dozen reviewers will give it a thumbs-down is $(.9)^{12}$, or .28, and hence the probability that at least one of the twelve will like the book is $1 - .28$, or .72. Thus even for a "bad" book, the chances of garnering a few favorable reviews grows rapidly with the number of reviewers, which fact, together with the painstaking excerpting of lukewarm reviews, goes some way toward explaining all the "riveting insights," "engaging romps," and "searing portrayals" on book jackets.

Of course, events are often dependent; the occurrence of one of them makes the occurrence of the other more or less likely. Having rolled a 6 with the first of a pair of dice, the chances of achieving a sum of 10, 11, or 12 with the second are greater than they otherwise would be. Knowing that someone is over 6 feet tall decreases the probability that that person is under 120 pounds. If a neighborhood contains a large number of Mercedeses, it probably does not contain many homeless people. These pairs of events are all dependent.

What we wish to ascertain in these cases is the conditional probability that one of these events has occurred or will occur given that the other has occurred or will occur. The conditional probability that the sum of the dice is 10, 11, or 12 given that the first die rolled comes up 6 is 1/2. There are six equally likely possibilities [(6,1), (6,2), (6,3), (6,4), (6,5), (6,6)] and three of them yield a sum of 10 or more. The conditional probability that one weighs less than 120 pounds given that one is over 6 feet tall is, I would estimate, not larger than 5%. This is considerably less than the probability of a randomly chosen person weighing in at under 120 pounds.

One must be careful when dealing with conditional probabilities. Note, for example, that the conditional probability that one can speak Spanish given that one is a citizen of Spain is approximately 95%, whereas the conditional probability that one is a Spanish citizen given that one can speak Spanish is not much more than 10%. Or consider the following scenario, a clarification of one taken from my book *Innumeracy* and about which I received a great many letters. It's known that in a certain curiously "normal" 1950s neighborhood every home houses a family of four—mother, father, and two children. One picks a house at random, rings the bell, and is greeted by a girl. (We assume that in the 1950s a girl, *if* there is at least one, will always answer the door.) Given these assumptions, what is the conditional probability that this family has both a son and a daughter? The perhaps surprising answer is not 1/2, but 2/3. There are three equally likely possibilities—older boy, younger girl; older girl, younger boy; older girl, younger girl—and in two of them the family has a son. The fourth possibility—older boy, younger boy—is ruled out by the fact that a girl answered the door.

Reckoning probabilities of complex events is generally not difficult *if* we've been given the probabilities of the constituent simple events. We can use Kolmogorov's axioms (involving the probability of mutually exclusive events, independent events, etc.), break the events down into mutually exclusive subevents, and calculate away. Or if this proves to be too troublesome, we can simulate the situation on a computer and determine the answer empirically (see the entry on *Monte Carlo method of simulation*).

Assigning probabilities to the basic events is a considerably more perplexing endeavor, however. A problem is that people's perceptions of crime or disease, to cite just two common examples, are shaped more by the dramatic portrayals of such on the eleven o'clock "action news" shows than they are by crime or health statistics. Add up the probabilities that any of the 5 billion-odd people in the world will kill you. The sum, depressingly high in this country, is still less than the probability that you'll kill yourself. Or consider that the average person in the United States is a quarter of a million times more likely to die from heart disease than from botulism. Needless to say, murder and an instance of botulism will make the news every time; suicide and a fatal heart attack will not (unless, of course, a celebrity is involved). The problem is not merely academic. An inability to assess and put into comprehensible perspective the hazards we face generally leads to unfounded and disabling personal anxieties or to unattainable and economically paralyzing demands for a risk-free environment.

Still, even when we calculate and estimate probabilities in the most idealized of situations, the philosophical question remains: What is probability?

THE PYTHAGOREAN THEOREM

∞

The ancient Greek mathematician Pythagoras (about 540 B.C.) and his predecessor Thales (about 585 B.C.) were arguably the first mathematicians ever. There were, of course, earlier peoples who possessed significant mathematical knowledge (the Rhind papyrus from 1650 B.C. is a most impressive lode of computational tools, including the clever use of a rudimentary notation for fractions), but the Egyptians, Babylonians, and others had a much different attitude toward the subject. Mathematics was for them only a practical subject useful for fixing taxes, figuring interest, determining the number of bushels of grain needed to make a quantity of beer, calculating areas of fields, volumes of solids, quantities of bricks, and astronomical particulars. These are without doubt extremely important skills that sadly are beyond the abilities of too many contemporary Americans, but since the time of Thales and Pythagoras mathematics has meant more than computation.

No one prior to the sixth century B.C. thought of mathematics as having a logical structure, as being capable of a rational systematization, as a set of ideal notions that might be clarified by the application of human reason. Thales, Pythagoras, and their contemporaries and followers did. No one before their time thought of numbers and geometric forms as ubiquitous, or conceived of theoretical circles and abstract numbers rather than of specific cart wheels and particular numbers. Thales and Pythagoras did. No one thought of isolating the more basic and obvious facts about these mathematical notions and then trying to derive from these fundamental facts other, less self-evident theorems by means of logic alone. Chorus: Thales and Pythagoras did. They and the Greek mathematicians who followed them invented mathematics (and logic) as we know it; they established it as a liberal art and not merely as practical number crunching.

Few personal facts about Pythagoras have survived. He traveled widely, established the mystical Pythagorean society which prohibited the eating of beans and whose motto was "All is number," and is said by some to have coined the words "philosophy" ("love of wisdom") and "mathematics" ("matters learned"). Pythagoras and his followers had an enormous influence on Greek mathematics (i.e., on mathematics) and are reputed to have contributed much of what 250 years later became the first two books of Euclid's *Elements,* in particular the theorem invariably attached to his name. (See the entry on *non-Euclidean geometry.*)

The Pythagorean theorem is a significant and indispensable one, and its standard proofs have been exemplars of geometrical beauty for almost two and a half millennia. It states that in any right triangle, the square on the hypotenuse (the side opposite the right angle) is equal in area to the sum of the squares on the other two sides. More symbolically put, if the two sides are of length A and length B and the hypotenuse is of length C, then we can be sure that $C^2 = A^2 + B^2$. [A Delphic hint for one proof of the theorem: Place four identical right triangles with sides A, B, and C into a very large square of side $(A + B)$ in two different ways. One placement should leave a region of area C^2 not covered by the triangles; another should leave two regions, one of area A^2 and the other of area B^2, not covered by the triangles.]

Although Pythagoras probably would have had little interest in doing so, one can use the theorem to compute distances. Thus, if Myrtle is 12 miles due north of the Parthenon and Waldo is 5 miles due east of the same edifice, then we can calculate that, as the crow flies, Myrtle is exactly 13 miles away from Waldo, since $5^2 + 12^2 = 13^2$. The length of a rectangle's or a shoebox's diagonal may be determined in a similar manner. Expressed in the language of analytic geometry (which was not to be discovered for 2,000 years) and greatly generalized, the Pythagorean theorem is a most powerful mathematical tool.

Nevertheless, Pythagoras would have better understood the aesthetic reaction of the philosopher Thomas Hobbes to his theorem. John Aubrey, Hobbes's friend, wrote that the philosopher "was 40 years old before he looked on geometry; which happened accidently. Being in a gentleman's library Euclid's *Elements* lay open, and 'twas the Theorem of Pythagoras. Hobbes read the proposition. 'By G——,' sayd he. (He would now and then sweare, by way of emphasis.) 'By G——,' sayd he, 'this is impossible!' So he reads the demonstration of it, which referred him back to such a proposition; which proposition he read. That referred him back to another, which he also read. *Et sic deinceps,* that at last he was demonstratively convinced of the trueth. This made him in love with geometry."

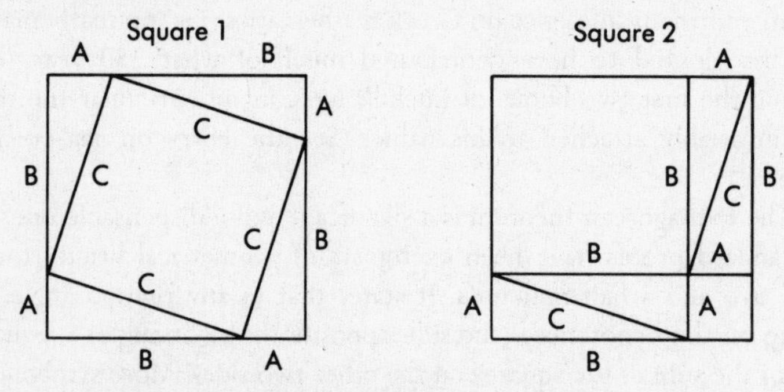

The area of square 1 equals the area of square 2.
The area of the four triangles in square 1 equals the area of the (same) four triangles in square 2.
Therefore the remaining area of square 1 is equal to the remaining area of square 2.
That is, $C^2 = A^2 + B^2$.

I must confess that I was moved by the same sort of infatuation when I first studied geometry (although I resisted the temptation to keep repeating "By G——"). Unfortunately, the invaluable legacy that is the axiomatic method, this derivation of nonintuitive propositions from self-evident axioms, is not being communicated to students; too many geometry texts seem to have reverted to a pre-Greek approach to the subject with their almost exclusive emphasis on unconnected facts, rules of thumb, and practical formulas. Pythagoras would have rather eaten beans than read some of these texts.

QED, PROOFS, AND THEOREMS

∞

A theorem is a proposition that follows by logic alone from the accepted axioms and other previously proved propositions. Usually only statements or propositions that are important or principal ones are given the honorific title theorem. An immediate consequence of a theorem is called a corollary to the theorem, while a lemma is a statement, usually technical, that is proved preparatory to the proof of the theorem. Pictures, diagrams, and examples may make a statement plausible, but only a detailed proof makes it a theorem.

Of course, this is only the official story. The author of a theorem in a mathematics research journal, a specialist in, say hemi-semi-demi-operator groups of prime power order, has typically sketched an argument that convinces himself, a couple of other hemi-semi-demi experts, and the referee. The result (mathematicians frequently call theorems "results") is most likely valid, but you might not want to bet your house on it.

Germane is a kind of experience I've had at a number of seminars, colloquia, and conferences. The speaker has filled the blackboard or his overhead transparencies with a dense barrage of definitions, equations, and proofs. I'm lost, but I notice that a good number of people in the audience are nodding sagely. During a break in the talk while the blackboard is being erased or the transparencies shuffled, I ask the enthusiastic nodder sitting next to me what a crucial term or symbol means. His sheepish shrug makes it clear that he's just as lost as I. The lecture begins and he resumes his nodding. I note that in addition to the nodders there are many nodders-off. There are also, I presume, a few mathematicians whose specialties are sufficiently close to the speaker's so that they don't feel the need to nod or the temptation to nod off. They are the temporary guardians of mathematical virtue.

In any case, to emphasize that a statement had indeed achieved the exalted status of theoremhood, the letters QED were traditionally placed at the end of its proof. An abbreviation for the Latin phrase "Quod erat demonstrandum," which means "That which was to be demonstrated," the letters sometimes served another purpose as well: intimidation. Responding questioningly to these three letters, which were capitalized in print and delivered with a flourish when spoken, took more self-confidence than most people could muster, mathematical diffidence being in all ages a fairly widespread condition. (Actually one needn't be innumerate or mathematically ignorant to be intimidated in this way. The words "It's trivial" dismissively applied to a theorem's nonexistent proof by an eminent mathematician often have much the same cowing effect on graduate students and even on professional mathematicians.)

Most texts now use a more low-key and functional dark vertical mark, ▮, at the end of a proof to indicate its conclusion. The practice, begun by the American mathematician Paul Halmos, is preferable since it is as serviceable as QED in marking closure, yet does not lend itself to intimidation. How, after all, do you say ▮ with a flourish? Nevertheless, I maintain that QED should still be employed occasionally for important theorems since the locution confers on the prover a more imperial feeling of satisfaction and finality than does the slightly plebeian ▮. There is, after all, a limit to how far one can go to avoid intimidation.

Mathematical logic has changed enormously over the past 2,500 years. Aristotelian syllogisms led to medieval classifications of arguments, which led in turn to Boolean algebras of propositions. Late-nineteenth- and twentieth-century logicians such as Frege, Peano, Hilbert, Russell, and Gödel have rigorized and vastly extended classical and medieval logic and created the powerful apparatus of modern predicate logic. Still, the essence of logic and the stick-it-in-your-eye appeal of mathematical proof remain the same and are reflected in the three letters QED. They signify ever so succinctly that the theorem follows *necessarily* from the assumptions and (assuming it's done right) that no one and nothing can change this.

One last point about proofs. Many people feel that only proofs expressed in symbolic form and utilizing all the paraphernalia of formal logic are acceptable. More often than not, however, such proofs merely muddle matters; preferable by far is a clear, compelling argument in words. See, for example, the proof that 6 is the minimum number of guests necessary to ensure that at least 3 will know each other or at least 3 won't in the entry on *combinatorics,* or the proof of the mountain climber's fixed-point property in the entry on *topology*.

THE QUADRATIC AND OTHER FORMULAS

∞

The quadratic formula is one of the first real theorems proved in high school algebra. It enables one to solve certain equations easily and seems to define mathematics for many people who assume that the whole subject is a matter of plugging numbers into such formulas or perhaps of factoring polynomials. It isn't, of course, but having a formula to plug into or a polynomial to factor does seem to bestow a certain sense of competence on one.

Quadratic equations are equations, such as $X^2 - 4X - 21 = 0$ or $3X^2 + 7X - 2 = 0$, in which the variable is squared ("quad" for "square"). Many situations in physics, engineering, and elsewhere lead to such equations. For example, if you're ever on a roof which is 200 feet off the ground and you throw a rock upward with a velocity of 80 feet per second, you might be interested in knowing that the rock will hit the ground after T seconds, where T is a solution to $-16T^2 + 80T + 200 = 0$ (the 16 is due to the effect of gravity).

What is required in these situations is to find the roots of these equations, those numbers which when substituted for X (or T) make the equations true statements. In the first equation above, the roots are -3 and 7; in the second the roots are approximately -2.59 and .26; and in the third, -1.83 and 6.83. There are various techniques (including guessing, factoring, and throwing rocks off of roofs) which enable one to find these roots, but none is more generally effective than the famous quadratic formula $X = \frac{-B \pm \sqrt{B^2 - 4AC}}{2A}$. (See the entry on *imaginary and negative numbers.*) The formula expresses the two roots of $AX^2 + BX + C = 0$ in terms of the numbers A, B, and C, which are equal to 1, −4, and −21, to 3, 7, and −2, and to −16, 80, and 200 in the three equations above. By plugging these numbers into the formula, the roots of each equation can be seen (at least after the haze of computation lifts) to equal the values noted.

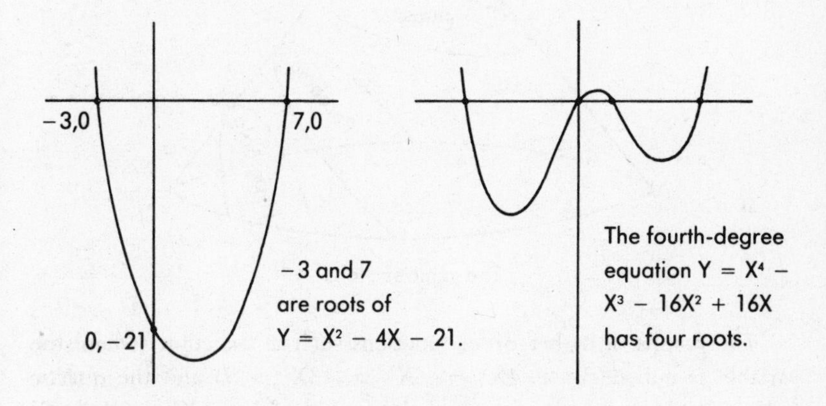

Often we're concerned not only with those numbers, if any, which result in a quadratic expression equaling 0 but also with the values of the expression for an arbitrary X. Thus, if the cost C required to produce X items is given by $100 + 2X + .1X^2$ ($100 in fixed cost, $2 per item made, plus a squared factor, $.1X^2$, which starts small but grows quickly as X does, reflecting storage and clerical costs associated with large inventories), we're led to the quadratic function $C = 100 + 2X + .1X^2$ and its graph. Or, if we're interested in the relation between $X^2 - 4X - 21$ and an arbitrary number Y, we'd study the function $Y = X^2 - 4X - 21$ and its graph. Typical of those of quadratic equations, these graphs are parabolic in shape, the graph of the latter function crossing the x-axis at −3 and 7 where $Y = 0$.

Examining the graphs of equations of the form $AX^2 + BXY + CY^2 + DX + EY + F = 0$, general quadratic equations in two variables, we meet the modern versions of Apollonius' ancient conic sections. Different values of the numbers A, B, C, D, E, and F give rise to equations whose graphs are circles, ellipses, parabolas, and hyperbolas. These are precisely the same figures that are formed by the intersection of a cone and a plane, the angle of the plane determining which one of the conic sections results.

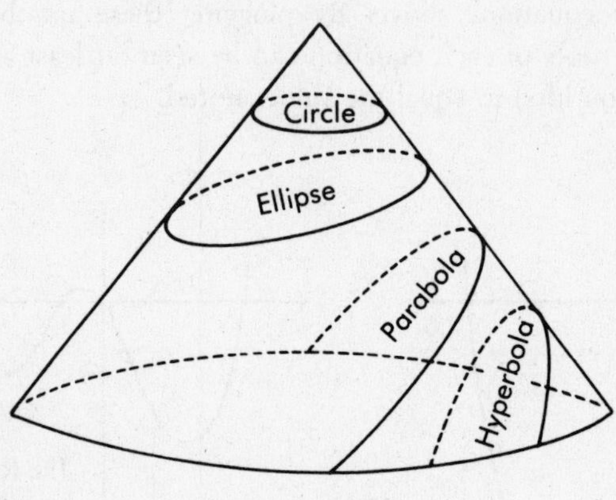

The conic sections

The graphs of higher-order functions such as the cubic (where the variable is cubed: $Y = 4X^3 - 5X^2 + 13X - 7$) and the quartic (where we have fourth powers of the variable: $Y = 6X^4 + 5X - 9$) are not as geometrically natural. As in the case of the quadratics, however, one can find the roots of these equations via formulas, but this time more difficult ones due to Geronimo Cardano, Niccolò Tartaglia, and other sixteenth-century Italian mathematicians. It's natural to assume therefore that all such higher-order equations (quintic, sextic, even those filthy septic ones) can be solved in the same way.

The nineteenth-century French mathematician Evariste Galois demonstrated, however, that this was not possible: Not all algebraic equations can be solved simply by plugging into an appropriate formula involving addition, subtraction, multiplication, division, and the taking of powers and roots. For higher-order equations such formulas don't exist. Galois approached the problem abstractly, focusing not on the roots of equations (or polynomial functions) but on the structures consisting of various permutations of the roots. (A not too misleading analogy would be to study families and their crises not by examining individual family members but by investigating these members' relations and dynamics.) The mathematicians of the day found his ideas incomprehensible, but these ideas have long since become part of the foundation of abstract algebra and provide one of the links between the latter subject and the one studied in high school.

QUANTIFIERS IN LOGIC

∞

Is fooling all of the people some of the time the same as fooling some of the people all of the time? To resolve this searing issue, we must focus on the meanings of "all" and "some." Elementary propositional logic (see the entry on *tautologies*) may be thought of as the thorough study of certain basic logical words: "if . . . , then . . . ," "if and only if," "and," "or," and "not." If we add to this list "every," "there is," "all," and "some," we have what is known as predicate logic, a logical system within which all mathematical reasoning may be formalized. These latter words are generally termed quantifiers (although they should perhaps be called qualifiers since they don't do much quantifying) and are used to transform relational forms involving variables into declarative statements. Thus, the form "X is bald" may be universally quantified to read "Every X is bald" or existentially quantified as "Some X is bald."

Consider the misanthropic form "X hates Y" in which each of the quantifiers may be quantified independently. If both variables are universally quantified, it translates as "For all X, for all Y, X hates Y," or, more naturally, "Everybody hates everyone." If the first variable is universally quantified and the second existentially quantified, we have "For all X, there is a Y (such that) X hates Y," or "Everybody hates someone." Switching the order of the quantifiers in the previous sentence gives us "There is a Y (such that) for all X, X hates Y" or, more colloquially put, "There is someone who is universally hated." If the first variable is existentially and the second universally quantified, the result is "There is an X (such that) for all Y, X hates Y," which in slightly better English is "There's somebody who hates everyone (himself and everyone else)." If both variables are existentially quantified, we obtain "There is an X, there is a Y (such that) X hates Y," or "Someone hates someone."

(The symbol for "for all" or "every" is an inverted A, that for "there exists" or "some" a backwards E. Thus, if we symbolize "X hates Y" by "H(X,Y)," then "Everyone hates someone" is symbolically rendered "$\forall X \exists Y$ H(X,Y)," while "There is someone who is universally hated" and "There's somebody who hates himself and everyone else" are translated into predicate logic as "$\exists Y \forall X$ H(X,Y)" and "$\exists X \forall Y$ H(X,Y)," respectively. How would you express "Everybody is hated by someone"?)

Formal manipulation of quantifiers is especially important in mathematics where there may be a string of several quantifiers in a row and where there is an enormous difference between "For every X there is a Y such that for all Z . . ." and "There is a Z such that for all X there is a Y . . ." A misplaced quantifier can change a continuous function into a uniformly continuous one, if you know what I mean (or even if you don't know what I mean). Such juggling is slightly less important in everyday life, where knowledge of a statement's context prevents errors from propagating. It's unlikely, for example, that "For every X, there is a Y such that Y is X's mother" will be confused with "For every Y, there is an X such that Y is X's mother." The latter states that everyone is a mother, while the former is the truism that everyone has a mother. Nor will many give credence to other permutations of the quantifiers: "There is a Y such that for all X, Y is X's mother" ("There is some person who is the mother of us all") and "There is an X such that for all Y, Y is X's mother" ("Some person is such that everybody is his or her mother").

One situation in which people are often sloppy about quantifiers in daily life is in the making of denials. George tells you that every inhabitant of the island is over 6 feet tall. If you want to deny this, you merely claim that someone on the island is under 6 feet tall, not necessarily that everybody is under 6 feet tall. To deny that there is someone every one of whose teeth has gold in it is to assert that everyone has at least one tooth that lacks gold.

English, unlike mathematics, is often ambiguous, and translation from an English sentence, especially a metaphoric one, to its formalization is often tricky. "All that glitters is not gold," for example, has two quite distinct formalizations, as does "Everybody's eaten" when uttered by cannibals. Even the simple English connective "is" may be translated into logic in very different ways. Compare the following: "Estragon is Mr. Beckett," where the "is" is the "is" of identity—$E = B$; "Estragon is anxious," where the "is" is the "is" of predication—E has the property A, or $A(E)$; "Man is anxious," where the "is" is the "is" of inclusion—For all X, if X has the property of being a man, then X has the property of being anxious, symbolically: $\forall X[M(X) \rightarrow A(X)]$; and "There is an anxious man here," where the "is" is existential (in the logical sense)—$\exists X[A(X) \wedge M(X)]$.

Returning to our original question, note that to fool some of the people all of the time is to take advantage of a particularly gullible group who are always certain to be bamboozled, while fooling all of the people some of the time is to keep up one's conning ways secure in the realization that eventually everyone will succumb to them. Finally, the way to formally express "Everybody is hated by someone" is "$\forall Y \exists X\ H(X,Y)$."

RATIONAL AND IRRATIONAL NUMBERS

∞

If you're not a mathematician or involved in mathematics, the definition of rational and irrational numbers won't impress you much—at least at first. A rational number is one that may be expressed as a ratio of two whole numbers (as fractions are); an irrational number cannot be so expressed. Most numbers one comes across in everyday life are rational: 3, which may be written as 3/1, 82 written as 82/1, 4 1/2 as 9/2, −17 1/4 as −69/4, 35.28 as 3528/100, or .0089 as 89/10,000. Moreover, between any two rational numbers, no matter how close, there are more. If people were to think about it, they would likely conclude that the "rational" in "rational number" was as redundant as the "fictional" in "fictional novel." Like life, however, rational numbers float in a sea of irrationality, and in an important and well-defined sense due to the mathematician Georg Cantor (see the entry on *infinite sets*), there are many more irrational numbers than rational ones. All these numbers, rational and irrational alike, are termed real numbers and may be expressed as decimals and arrayed along a line, which is known, appropriately enough, as the number line.

The first irrational number discovered was $\sqrt{2}$, the square root of 2 (that number which when multiplied by itself equals exactly 2), and the discovery of its irrationality led to something of a crisis in early Greek mathematics. The standard indirect proof of $\sqrt{2}$'s irrationality is, like the proof of the infinitude of prime numbers, so elegant and so illustrative of the classical reductio ad absurdum technique that I include it here. Assume that, contrary to our beliefs, $\sqrt{2}$ is rational and is equal to the ratio of P to Q, P/Q. Now cancel any common factors in P and Q to reduce the fraction to lowest terms. For example, if P/Q were 6/4, we would rewrite it as 3/2. I'll indicate this reduced fraction as M/N, where M and N are whole numbers having no factors in common.

Squaring both sides of the equation $\sqrt{2} = M/N$, we obtain $2 = M^2/N^2$, which, on multiplying by N^2, becomes $2N^2 = M^2$. Now, as we shall see, we're led ineluctably from this equation to an absurd conclusion (the reductio ad absurdum) that demonstrates the untenability of our assumption of $\sqrt{2}$'s rationality. This feeling of ineluctability, I think, provides a good deal of the psychological payoff for doing mathematics and should be appreciated even if the details of the proof are not.

Proceeding without any more interruptions, we observe that the left side of the equation $2N^2 = M^2$ contains a 2 as a factor and so it must be even. Hence the right side must be even too. Since M^2 is even, M must also be even, since the square of an odd number is itself odd. Thus M, being even, equals 2K for some whole number K, and we have $M^2 = (2K)^2 = 4K^2$. Substituting $4K^2$ into the earlier equation, we have $2N^2 = 4K^2$, or upon division by 2, $N^2 = 2K^2$. Since the right side of this last equation contains a 2 as a factor and is thus even, so too is N^2 and hence N must be even as well, since the square of an odd number is odd. Since N is even, it must equal 2J for some whole number J. We insisted at the outset that M and N have no common factor, but M and N, as we've shown, have the factor 2 in common, M being equal to 2K and N to 2J. This contradiction is a direct consequence of our original assumption that $\sqrt{2}$ is rational and thus makes that assumption untenable. We conclude that $\sqrt{2}$ must be irrational. That's it. QED. Trumpets resound.

We can demonstrate the irrationality of many other numbers. The product of two rational numbers is rational, so now that we know $\sqrt{2}$ is irrational, we know that $\sqrt{2}/2$ must be irrational as well. Otherwise $2 \times \sqrt{2}/2$ (which equals $\sqrt{2}$) would be rational and we just proved it wasn't. Likewise $\sqrt{2}/3$, $\sqrt{2}/4$, $\sqrt{2}/5$, and so on are all irrational. Other algebraic numbers such as the square root of 3, the cube root of 5, and the seventh root of 11 are also irrational, as are π, e, and uncountable hordes of anonymous numbers. (For computational purposes, the rational approximation to these irrational numbers is sufficient. We may use 1.4 or 14/10 to approximate √2, or, if more accuracy is needed, we would use 1.41 or 1.414.)

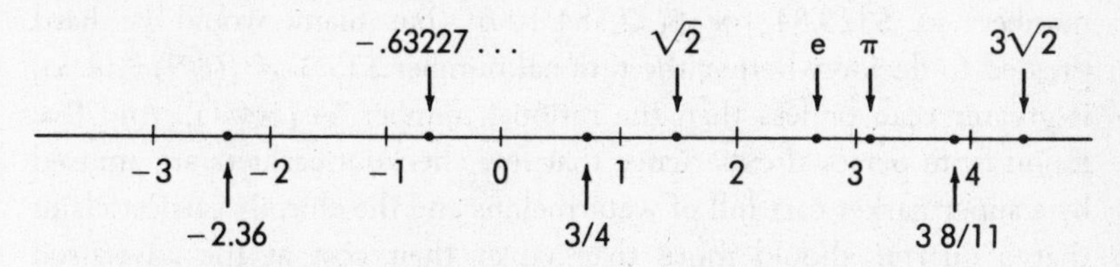

The number line with a few rational numbers listed below and a few irrationals above it

When we express √2 (whose decimal expansion continues 1.4142135623730950488016887242096980785697 . . .) and other irrationals as decimals, we find that there is no repeating pattern of digits in their endless decimal expansions. Rational numbers, by contrast, all eventually have sequences of digits which repeat. The decimals 5.3333 . . . , 13.8750000 . . . , 29.38 461538 461538 461538 . . . , representing 5 1/3, 13 7/8, and 29 5/13, respectively, are all rational numbers with repeating decimals. Since any rational number may be expressed as a quotient of whole numbers, the reason for this repetition is clear. The first remainder in the division process may be only one of finitely many numbers and thus continuing the process inevitably results in the same remainder appearing again and hence to a repetitive pattern in the decimal expansion of the rational number. The converse is trickier, but it can be shown that if there is a repetitive pattern to the decimal expansion of a number, then the number is the sum of an infinite geometric series which always turns out to be rational.

I reiterate that a number is rational if and only if its decimal expansion is ultimately repetitive. The decimal expansions of √2 and π and e display no such repetition. In the set of all possible decimal expansions (which is to say: in the set of all real numbers), pattern and repetition are much rarer than their absence. Harmony is always much rarer than cacophony.

In closing, I must note that despite being rarer, rational numbers play a larger role in narrowly practical affairs than do irrational ones. In most daily and business contexts, the distinction between them isn't nearly as important as is an ability to deal comfortably with rational numbers in their various incarnations: fractions—proper, improper, and mixed, decimals, and percentages. Unfortunately this ability isn't as universal as it should be. Most people can handle such pleasing rational numbers as \$325.84 [or \$(32,584/100)], but many would be hard pressed to decide whether the rational number \$25/3 × [(8/9) − (2/5)] is greater than or less than the rational number \$4 [\$(4/1)]. And I've found from personal experience that few checkout cashiers are amused by a supermarket cart full of watermelons and the chirpily earnest claim that a quarter should more than cover their cost at the advertised bargain price of .59¢ per pound [(59/100)th of a cent, (59/10,000)th of a dollar].

RECURSION—FROM DEFINITIONS TO LIFE

∞

The expression 5! indicates the product 5 × 4 × 3 × 2 × 1, 19! the product 19 × 18 × 17 × . . . × 3 × 2 × 1. These locutions are read "5 factorial" and "19 factorial," respectively, not "five!" or "nineteen!" with an exclamatory flourish. Although their primary use is in probability and other areas of mathematics where counting up possibilities is important, I sometimes think that it would be beneficial if people thought of each other as "historical factorials." Thus, (Myrtle)! would be understood not just as present-day Myrtle but as the product of all her past experiences.

Formally, we stipulate that 1! is 1 and then define (N + 1)! to be equal to (N + 1) × N!. That is, we define "factorial" explicitly for the first term and then define its value for any other term via its values for the predecessors of that term. A definition of this sort is called recursive, and we can use it to calculate the value of 5!. It's 5 times 4!. But what is 4!? It's 4 times 3!. What is 3!? It's 3 times 2!. What is 2!? It's 2 times 1!. And finally the definition tells us what 1! is. It's 1. Putting this all together, we determine that 5! = 5 × 4 × 3 × 2 × 1. Happily I won't go through this litany for 19!.

In a similar manner the recursive definition of addition tells us how to add any number to X. It stipulates that X + 0 is X, and recursively defines X + (Y + 1) to be equal to 1 more than X + Y. (My apologies if the rest of this paragraph reads like a numerical shaggy dog story.)

To determine (8 + 3) using the above definition, for example, we note that 8 + 3 = (8 + 2) + 1, that 8 + 2 = (8 + 1) + 1, that 8 + 1 = (8 + 0) + 1, and that (8 + 0) = 8. Putting these all together, we get that 8 + 3 = 8 + 1 + 1 + 1 = 11. We've reduced addition to counting. The recursive definition of multiplication stipulates that X × 0 = 0 and then defines X × (Y + 1) to be (X × Y) plus X. By tracing through the definition we can determine the value of 23 × 9 by reducing it to a series of additions [23 × 9 = (23 × 8) + 23; and (23 × 8) = (23 × 7) + 23; and (23 × 7) = (23 × 6) + 23, and . . .], which in turn are reducible to countings. Likewise, we can say what it means to raise X to any power. We stipulate that X^0 is 1 and define $X^{(Y+1)}$ to be X times X^Y. Hence, $7^4 = 7 \times 7^3$, $7^3 = 7 \times 7^2$, and so on—exponentiation reduced to multiplication reduced to adding reduced to counting.

Although it may seem pointless at first, the idea of recursively expressing the value of a function at (N + 1) in terms of its predecessors' values is a very powerful one and is indispensable to computer science. In fact, with its characteristic employment of loops (the performing of some procedure again and again for various values of some variable), subroutines, and other strategies for reducing complex procedures to simple arithmetical operations, recursion is at the very heart of computer programming. The mathematical functions and algorithms that can be defined in a recursive way turn out to be precisely the ones that computers can deal with. That is, if a function is recursive, a computer can calculate it, and if a computer can calculate a function, then that function is recursive. Furthermore, these recursive definitions can be nested and iterated indefinitely and, via appropriate codings and correspondences, can be extended to all sorts of activities that seem not to have much to do with computation.

These functions and definitions play an important role in logic since they make precise what one means by words like "mechanical," "rule," "algorithm," and "proof" (see also the entry on *mathematical induction*). In formal grammar, linguists use recursive definitions to clarify the rules of grammar and study cognitive processes. They demonstrate how long complex statements may be built up recursively from short clauses and phrases. Combined with self-reference, recursion is even more powerful. Some computer viruses, for example, reproduce in something like the manner of the following sentence, which provides directions and raw material for its own replication. *Alphabetize and append, copied in quotes, these words: "these append, in Alphabetize and words: quotes, copied."* Less concisely put, the previous sentence directs that the words following the colon be alphabetized and then to this alphabetized list should be appended the unalphabetized words in quotes. Presto! The sentence has reproduced itself and its descendants will do the same.

Recursion also plays an increasingly significant role in describing physical phenomena, especially with the development of chaos theory, which makes vivid how complicated and lifelike recursively defined structures can be.

A marked square and its 8 neighbors

From one generation to the next: Marked squares having 2 or 3 marked neighbors stay on. Those with 0, 1, or more than 4 marked neighbors die out. An empty square having exactly 3 marked neighbors comes to life.

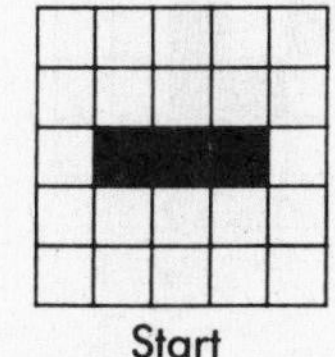

Start

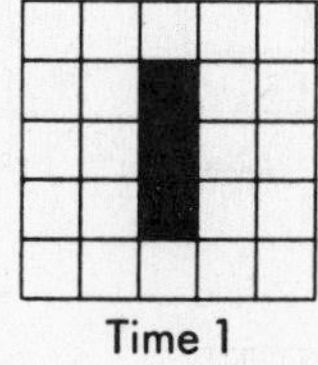

Time 1

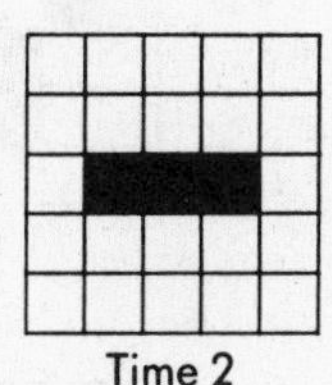

Time 2

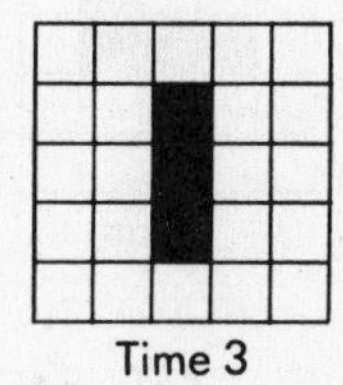

Time 3

A blinking pattern

In this last regard, consider a fascinating application of a simple recursive definition. British mathematician John Conway's solitaire game "Life" takes place on an infinite checkerboard some of whose squares are occupied by markers. (Graph paper with some of its squares darkened works just as well.) Since every square on the board has 8 neighboring squares (4 adjacent and 4 diagonal), each marker may have anywhere from 0 to 8 neighbors. The original distribution of markers is the first generation, and passing from one generation to the next is regulated by the following three rules. Every marker having 2 or 3 neighbors stays on the board and continues into the next generation.

Every marker with 4 or more neighbors is removed from the board in the next generation, as are those with 0 or 1 neighbor. Each empty square having exactly 3 occupied neighbors has a marker placed on it in the next generation.

The changes dictated by the three rules all take place simultaneously, discrete ticks of the clock bringing about the succeeding generations of this "cellular automaton." Surprisingly, these simple recursive rules governing the survival, death, and birth of markers lead to unpredictably complicated and beautiful patterns and movements across the checkerboard. Figures resembling trains and planes shoot across the board as succeeding generations are produced. Some initial configurations die out, others reproduce, while still others spawn whole universes of toads and ships and geometrical designs. Some oscillate or blink, others cycle through a sequence of change, and still others never stop evolving.

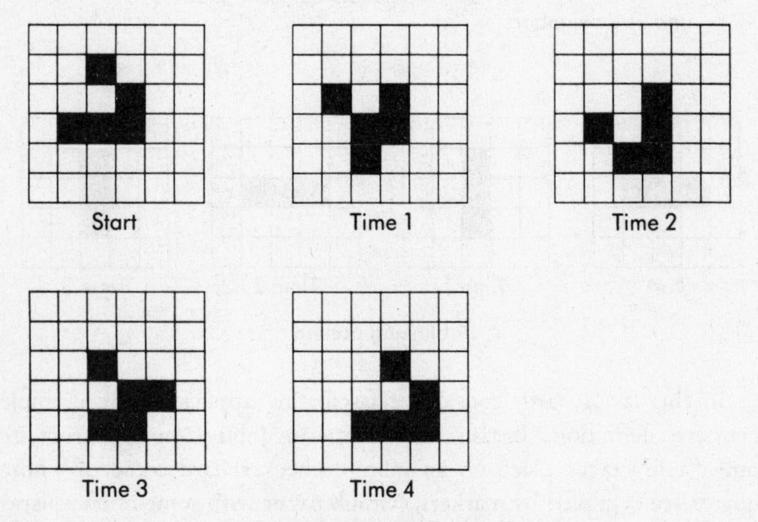

A pattern reproduces itself.

To get a feel for these changing patterns it's necessary to either start with some markers on a checkerboard (or with darkened squares on graph paper) and systematically follow the rules by hand or else watch a computer presentation of the game. Try various initial distributions of markers. You might experiment by first examining three markers in a row. This pattern is called a blinker, alternating between the horizontal and vertical. Try a few others. The pattern with 4 markers, 3 in a row horizontally and 1 more placed above the middle marker, has a particularly interesting development.

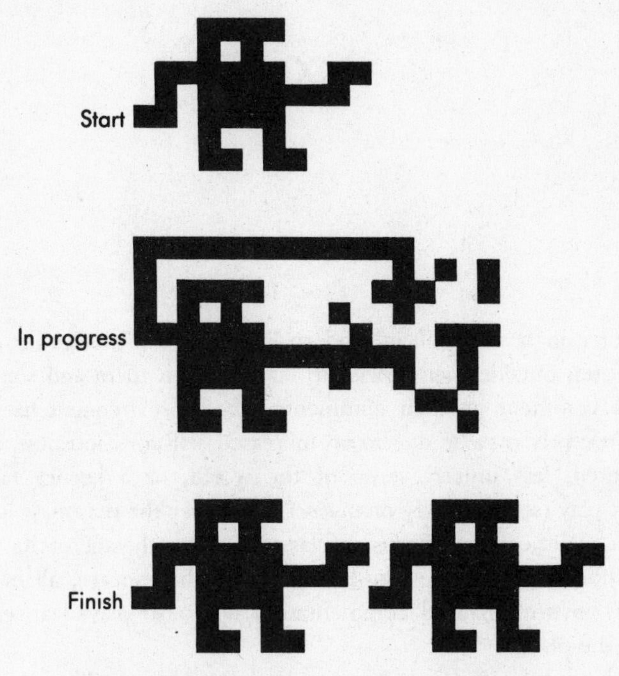

Simplified diagram of a pattern reproducing

Conway has even proved that certain initial configurations of markers may be employed as a slow, but full-fledged computer, which fact brings us back to recursive functions of integers. The essence of "Life" and possibly of life too is recursion (see the entry on *Turing's test*). Mathematicians from Pythagoras to Poincaré have written that numbers and counting are the basis of all mathematics. Recursive definitions give us a sense in which this claim is actually modest.

RUSSELL'S PARADOX

∞

It is increasingly common in modern literature and movies for characters to step outside their stories to comment on them and sometimes even to comment on their comments. The more frequent use of the device recently may be due to an increased self-consciousness, a more fragmented, less unitary sense of the world, or a keener taste for abstract play (see the entry on *humor*). Whatever the reason, it is a very old idea. Witness the chorus in classical Greek theater or its various incarnations through the Middle Ages and Shakespeare, all of which acted as institutionalized commentators who also played an essential part in the play.

Such multilayered commentary leads to more lifelike complexity and, as the classic liar's paradox already indicated over 2,000 years ago, sometimes to paradox. Epimenides the Cretan is reported to have stated that all Cretans are liars. The crux of his Janus-faced pronouncement is clearer if we simplify his statement to "I am lying" or, better yet, to "This sentence is false." Assigning the label Q to "This sentence is false," we notice that if Q is true, then by what it says it must be false. On the other hand, if Q is false, then what it says is true, and so Q must then be true. Hence, Q is true if and only if it is false. An attenuated form of the liar's paradox is present, albeit implicitly, whenever the frame of a painting, the stage setting of a theater production, or the tone of voice one uses to tell a joke suggests, "This is false, not real."

The self-referential aspect of these paradoxes can also be expressed in other ways. Consider the well-known case of the town barber who is legally required to shave all those men and only those men who do not shave themselves. The poor barber is left to wonder whether to shave himself or not. If he shaves himself, by law he should not. On the other hand, if he does not shave himself, by law he should. A version of the paradox that is closer to the focus of this entry deals with the mayoral residency laws of a certain country. Some of the mayors live in the cities they govern, while others do not. A decree is issued by an earnest reform-minded monarch who requires all nonresident mayors and only nonresident mayors to live in one place—call it city DMZ for Displaced Mayors Zone. Suddenly realizing that DMZ requires a mayor of its own, the monarch develops a royal headache pondering where DMZ's mayor should reside.

We arrive finally at Russell's paradox. Due to the English philosopher and mathematician Bertrand Russell, it has something of the same flavor as the earlier paradoxes, but with considerably more mathematical potency. Although it's framed in terms of set theory, the only thing you need know about sets is that, informally at least, they are well-defined collections of objects of any sort whatever. One bit of notation is also helpful. To indicate that 7 is a member of P, the set of all prime numbers, or that 10 is not a member of P, we write $7 \in P$ and $10 \notin P$, respectively. Thus $13 \in P$ and $15 \notin P$. In general, "$X \in Y$" means that X is a member of the set Y or, equivalently, that Y contains X as a member. Hence, if Y is the set of countries in the United Nations and X is Kenya, then $X \in Y$.

Now, to derive the paradox we note that some sets contain themselves as members ($X \in X$) and some do not ($X \notin X$). The set of all things mentioned on this page is itself mentioned on this page and thus contains itself as a member. Likewise the set of all those sets with more than eleven members itself contains more than eleven members and thus is a member of itself. And certainly the set of all sets is itself a set and thus it too is a member of itself. Most naturally occurring sets do not contain themselves as members, however. The set of senators in the U.S. Senate is not itself a senator and thus not a member of itself. Similarly the set of odd numbers is not itself an odd number and thus does not contain itself as a member.

Let us denote by M the set of all those sets that contain themselves as members and by N the set of all those sets that do not contain themselves as members. More symbolically phrased, for any set X, we have $X \in M$ if and only if $X \in X$. On the other hand, for any X, $X \in N$ if and only if $X \notin X$. Now, we may ask whether N is a member of itself or not. (Compare this question with "Who shaves the barber?" or "Where does the mayor of city DMZ live?") If $N \in N$, then by definition $N \notin N$. But if $N \notin N$, then by definition $N \in N$. Thus N is a member of itself if and only if it is not a member of itself. This contradiction constitutes Russell's paradox.

Russell's resolution of this paradox (there are others) is to restrict the notion of a set to a well-defined collection of already existing sets. In their famous theory of types he and Alfred North Whitehead classified sets according to their type. On the lowest level, type 1, are individual objects. On the next level, type 2, are sets of type 1 objects. On the next level, type 3, are sets of type 1 or of type 2 objects, and so on. The elements of type N are sets of objects of type $(N - 1)$ or lower. In this way the paradox is avoided since a set can be a member only of a set of higher type and not of itself. A set's being a member of itself ($X \in X$) is ruled out, as are sets like M and N defined in terms of this notion.

(I should mention that Russell and Whitehead constructed the theory of types not only to prevent this paradox and others like it but, more importantly, to provide an axiomatic foundation for the whole of mathematics. They succeeded in reducing all of mathematics to logic as embodied in the theory of types—logic together with the above hierarchical notion of set. I also stress that self-reference need not and, in fact, doesn't usually lead to paradox in mathematics or in natural languages. Most instances of the word "I," for example, are entirely unproblematic.)

We may apply a type solution to the liar's paradox as well and require that "All Cretans are liars" be assigned a higher type than other statements made by Cretans. We make distinctions among first-order statements, which do not refer to other statements at all (e.g., "It's raining" or "Menelaus is bald"); second-order statements, which refer to first-order statements ("Waldo's remarks about the food were crazy"); third-order statements, which refer to second-order statements; and other higher-order statements. Thus, when Epimenides says "All Cretans are liars," he is to be interpreted as making a second-order statement that does not apply to itself but only to first-order statements.

If he states that all his second-order statements are false, that assertion is third-order and does not apply to itself.

More generally, the whole concept of truth can be given a level structure: $truth_1$ for first-order statements, $truth_2$ for second-order statements, and so forth. This notion of truth has been extensively developed and formalized by the mathematician Alfred Tarski and rendered more flexible for tangled natural languages by the philosopher Saul Kripke. One sort of entanglement occurs when two or more people each make unproblematic assertions which, taken together, yield paradox. A simple example is the exchange in which George claims, "What Martha says is false," and Martha states, "What George says is true," and neither says anything else.

What I find odd is that these matters are usually considered impossibly esoteric and academic by the same people who routinely unravel the most convoluted and multilayered of narratives—stories of teenage intrigue (she said that he said, but that can't be so or she wouldn't have said such and such when he lied to her about what he thought she said), labyrinthine analyses of Middle Eastern politics, or the movies and novels I mentioned at the beginning. (See also the entry on *variables.*)

SCIENTIFIC NOTATION

∞

Quick, which is bigger, 381000000000 or 98200000000? And which is smaller, .000000000034 or .00000000085? Without counting zeros or the presence of separating commas, more than a glance is needed to comprehend 9,450,000,000,000,000, the number of meters in a light-year, or .000 006 5, the wavelength of red light in meters. Scientific notation is a convention by which we can write these very large and very small numbers in such a way that their relative magnitudes are immediately clear. It also spares us the effort of having to pronounce them.

To understand the notation it's necessary to recall that 10^1 is another way to write 10, that 10^2 is 100, that 10^6 is 1,000,000, and that in general 10^N is 1 with N zeros following it. Hence 4×10^3 is $(4 \times 1{,}000)$ or 4,000, and 5.6×10^6 is $5.6 \times 1{,}000{,}000$ or 5,600,000. To indicate small numbers, negative exponents are used, and we define 10^{-1} to be .1, 10^{-2} to be .01, 10^{-6} to be .000 001, and in general 10^{-N} to be a 1 preceded by $(N - 1)$ zeros and a decimal point (alternatively, $10^{-N} = 1/10^N$). Hence 5.7×10^{-3} is $(5.7 \times .001)$ or .0057, and 9.1×10^{-8} is $(9.1 \times .000\ 000\ 01)$ or .000 000 091.

It's a part of the convention to have just one nonzero digit to the left of the decimal point. Thus 48,700 is written as 4.87×10^4 rather than as 48.7×10^3, and .000 000 23 is written as 2.3×10^{-7} rather than as 23×10^{-8}, although in each case both expressions indicate the same number. Likewise, we write 239,000,000 as 2.39×10^8; 59,700,000,000,000 as 5.97×10^{13}; .000 031 as 3.1×10^{-5}; and .000 000 002 5 as 2.5×10^{-9}.

In addition to ease of recognition, scientific notation makes rough order-of-magnitude calculations easier since we can make use of the law of exponents, which says that $10^M \times 10^N$ is equal to 10^{M+N} or, in particular, that $10^5 \times 10^8 = 10^{13}$. For example, since there are about 5×10^9 people in the world, and since each person has on the average approximately 1.5×10^5 hairs on his or her head, there are $(5 \times 10^9) \times (1.5 \times 10^5)$, or 7.5×10^{14}, human hairs in the world. (I digress to note that the law of exponents motivates the use of fractional exponents as well. If $X^? \times X^? = X^1$, then $X^?$ must equal $\sqrt{X}$, since squaring it equals X^1. But $X^{1/2} \times X^{1/2}$ also equals X^1 by a formal application of the law of exponents. Thus $X^{1/2}$ is defined to be $\sqrt{X}$.)

In order to translate the hairy 7.5×10^{14} figure (750 trillion) and others into English, one needs the scientific notation for certain traditional number words. A thousand is 10^3; a million is 10^6; a billion, 10^9; a trillion, 10^{12}. A visceral grasp of the differences between these numbers can be had if it's noted that a thousand seconds takes about 17 minutes to tick by, a million seconds requires about 11 1/2 days, a billion seconds approximately 32 years, and a trillion seconds almost 32,000 years.

Knowing these equivalences and a few common facts (the population of the United States, of the world, the distance from coast to coast) makes it easier to evaluate the reliability and significance of various large numbers and to assess risks in a rational manner. To cite a topical example, the number of commercial sexual acts performed by American prostitutes annually is an estimated 300 million (3×10^8 if you like). Is this reasonable, and if not, is the true number higher or lower? The answer is of interest not only to academics and adolescents, but also to AIDS epidemiologists trying to explain the seeming dearth of men whose disease can be attributed to such encounters. Or consider the (by some estimates) half-trillion-dollar (5×10^{11}) savings-and-loan scandal. Does anyone doubt that it stayed invisible for so long because it was a "numbers story" and not a "people story"?

Often prefixes are used rather than exponents. The most common are kilo- for 10^3, mega- for 10^6, giga- for 10^9, and tera- for 10^{12}. In the other direction the prefix milli- indicates 10^{-3}; micro-, 10^{-6}; nano-,

10^{-9}; and pico-, 10^{-12}. A nanosecond is a billionth of a second, and thus bears the same relation to a second as the latter does to the thirty-two years of a gigasecond.

Coined by mathematician Edward Kasner, the fanciful word "googol" leads us into even more fathomless realms by giving a name to the number 1 followed by a hundred zeros: 10^{100}. A googolplex is incomparably more inconceivable, being defined as a 1 with googol many zeros after it: $10^{(\text{googol})}$. It's hard to come by any sets of real objects that have this many members. The physicist Arthur Eddington once wrote that there were approximately 2.4×10^{79} particles (fewer than a googol) in the entire universe, and though the physics upon which this estimate was based is quite outmoded, the fact remains that if we stick to real entities these vast numbers are much more than adequate for any purpose. When, however, we start counting *possibilities,* it's easy to surpass even these numbers. If we flip a penny just 1,000 times, for example, the set of all possible sequences of heads and tails is, by the multiplication principle, $2^{1,000}$, or about 10^{300}, which is a googol cubed!

[Useful in some contexts, scientific notation is of course inappropriate in many others. At a special nursery school for "gifted" preschoolers in Philadelphia I once saw the academically obsessed parents of a diapered genius spread their arms wide and say to him, "We love you 10^8." Dressed like 1960s hippies, they were just clowning around, but I couldn't help wondering if they went home afterward and listened to the old folk song "I Will Live with You 3.65×10^2 Days" or told their kid about "Scheherazade's 10^3 or So Nights."]

SERIES—CONVERGENCE AND DIVERGENCE

∞

Infinite series and their applications constitute an important area of mathematical analysis. Used informally by mathematicians long before they were completely understood (see the entry on *Zeno*), series still appeal seductively to our intuitions about number and infinity. Speaking a little loosely (N symbolizing an arbitrary positive integer and . . . indicating that the series continues ad infinitum), I maintain that the sum $1 + 1/3 + 1/9 + 1/27 + 1/81 + 1/243 + 1/729 + \ldots + 1/3^N + \ldots$ is finite, whereas the sum $1 + 1/2 + 1/3 + 1/4 + 1/5 + 1/6 + \ldots + 1/N + \ldots$ is infinite. Why the difference, and so what?

The rough answer to the Why question is that the terms of the first series get small quickly enough to have a bounded sum, and those of the second series don't. In the first series one passes from one term to the next by dividing by 3. In the second the terms shrink very slowly and one can prove that if enough of them are added together, their sum will eventually exceed any number—say a billion trillion. The first series is said to converge (its sum is 3/2, or 1 1/2), and the second is said to diverge to infinity. (The verbs "converge" and "diverge" may sound a little odd here, but they are standard mathematical usage.)

It's not always obvious whether a series converges or diverges. The series $1 + 1/4 + 1/9 + 1/16 + 1/25 + \ldots + 1/N^2 + \ldots$ converges (wondrously its sum is $\pi^2/6$), while $1/\log(2) + 1/\log(3) +$

1/log(4) + . . . diverges. (For those who like notation, I note that the Greek letter Σ is frequently used to symbolize a series and ∞ to indicate infinity; thus $\sum_{N=1}^{\infty} \frac{1}{N^2} = \frac{\pi^2}{6}$ and $\sum_{N=2}^{\infty} \frac{1}{\log(N)}$ diverges.) The series whose terms contain successive factorials in the denominator, 1 + 1/1! + 1/2! + 1/3! + 1/4! + 1/5! . . . + 1/N! + . . . , converges (again wondrously, I think) to the number e, while the series 1/1,000 + 1/(1,000 × √2) + 1/(1,000 × √3) + 1/(1,000 × √4) + 1/(1,000 × √5) + . . . 1/(1,000 × √N) + . . . diverges.

What is meant by convergence of a series may be clarified by the notion of a "partial sum." The idea is to sneak up on the "infinite sum" by means of a sequence of partial sums. For the three convergent series mentioned so far, this means: The sequence of partial sums 1, 1 + 1/3, 1 + 1/3 + 1/9, 1 + 1/3 + 1/9 + 1/27, . . . gets arbitrarily close to 1 1/2; the partial sums 1, 1 + 1/4, 1 + 1/4 + 1/9, 1 + 1/4 + 1/9 + 1/16, . . . approach $\pi^2/6$ with arbitrarily fine precision; likewise, the differences between the partial sums 1, 1 + 1/1!, 1 + 1/1! + 1/2!, 1 + 1/1! + 1/2! + 1/3!, . . . and the mathematical constant e shrink to zero (see the entry on *e* for its definition).

Relatively easy to deal with are so-called geometric series in which each term of the series is obtained from its predecessor by multiplication by a constant factor. The first series above is an example, as is 12 + 12(1/5) + 12(1/25) + 12(1/125) + 12(1/625) + ... *Finite* geometric series arise in many everyday contexts. If each year one invests $1,000 at 10% interest in a college fund, then the fund's worth after 18 years will be $1,000 + $1,000(1.1) + $1,000(1.1)^2 + $1,000(1.1)^3 + $1,000(1.1)^4 + . . . + $1,000(1.1)^{18}, or about $56,000. The first term of the series is the $1,000 you just invested. The second term of $1,000(1.1), 110% of $1,000, is the value of the $1,000 you invested one year ago. The third term is the compounded value of the $1,000 you invested two years ago, and so on until the last term of the series, $1,000(1.1)^{18}, which is what the initial $1,000 you invested eighteen years ago is worth today.

Annuities provide a concrete example of an infinite geometric series. How much money would need to be deposited in an account today so that you, your heirs, and your heirs' heirs will be able to withdraw $1,000 per year from it forever? Assuming a constant interest rate of

10%, it's clear that only \$10,000 is required to generate \$1,000 in interest every year.

It may be a little more difficult to see that \$10,000 is what the following infinite series sums to: \$1,000 + \$1,000/1.1 + \$1,000/$(1.1)^2$ + \$1,000/$(1.1)^3$ + \$1,000/$(1.1)^4$ + . . . + \$1,000/$(1.1)^N$ + ... The first term in the series is the \$1,000 you withdraw tomorrow. The next term is the \$1,000 you withdraw next year (you *divide* it by 1.1 or 110% because its present value is that much less than \$1,000). The term after that is the \$1,000 you withdraw in two years [divided by $(1.1)^2$ because its present value is correspondingly even smaller than \$1,000]. And each succeeding term is divided by another factor of 1.1, reflecting another year's devaluation. If this annuity were to be awarded by lottery, it might be dubbed the infinite end-of-time lottery: \$1,000 a year forever and ever. It's less misleading to quote the \$10,000 present value. Likewise \$1 million a year forever is worth only about \$10 million.

Geometric series arise when we are concerned with determining the quantity of medicine in the blood of a person on a long-term daily regimen of the medicine. This is because we must sum up the amount in the blood from today, yesterday (a fraction of today's), the day before yesterday (the same fraction of yesterday's), and so on. For similar reasons such series also come up when one is interested in either the total impact of a purchase of government securities by the Federal Reserve Bank (the ripple effect) or in the total distance traveled by a bouncing ball. The formula for the sum of an infinite series of the form $A + AR + AR^2 + AR^3 + AR^4 + AR^5 + \ldots + AR^N + \ldots$ is A/(1-R). A is the initial term of the series and R is the constant factor by which terms are multiplied to get their successors. In the first geometric series above, A is 1 and R is 1/3, so the sum, 1/(1-R), is 1/[1 − (1/3)] or 3/2. What is the sum of 12 + 12(1/5) + 12(1/25) + 12(1/125) + 12(1/625) + . . . ?

Unfortunately not all series are geometric series and sophisticated rules and tests have been devised for determining convergence, for calculating the rate of convergence, for finding the sum, and so on. The sum of a series, it should be reiterated, is defined by considering the sequence formed by its partial sums. In the series 1 + 1/4 + 1/27 + 1/256 + . . . $1/N^N$ + . . . , this sequence is 1, (1 + 1/4), (1 + 1/4 + 1/27), (1 + 1/4 + 1/27 + 1/256), ... If this sequence approaches a limit, as this one does, that limit is said to be the sum of the series.

Alternatively phrased, if the sequence of partial sums gets arbitrarily close to some number, that number (in this case something less than 3/2) is said to be the sum of the series. (See the entry on *limits.*)

These rules and tests for convergence are especially valuable when dealing with power series (or infinite polynomials). Normal (finite) polynomials are algebraic expressions such as $3X - 4X^2 + 11X^3$, $7 - 17X^2 + 4.7X^5$, and $2X + 5X^3 - 2.81X^4 + 31X^9$. Since the algebra and calculus of such polynomial functions is especially easy, mathematicians have attempted to approximate other common functions with them. (See the entry on *functions.*) For a wide class of functions this is possible. It can be shown, for example, that the trigonometric function sin(X) may be represented by the power series (infinite polynomial) $X - X^3/3! + X^5/5! - X^7/7! + X^9/9! \ldots$ and approximated by a finite partial sum of this series. That is, sin(X) is approximately equal to $X - X^3/3! + X^5/5! - X^7/7!$. Likewise, the exponential function e^X may be represented by the series $1 + X + X^2/2! + X^3/3! + X^4/4! + \ldots X^N/N! + \ldots$ and approximated by the polynomial sum of the first few terms. Thus, the value of e^2 is approximately equal to $1 + 2 + 2^2/2! + 2^3/3! + 2^4/4! + 2^5/5!$.

Finding derivatives and integrals, solving differential equations, and working with complex and imaginary numbers are all considerably simplified if we're dealing with functions which, like e^X and sin(X), are representable by power series. It's difficult, in fact, to overestimate the importance of infinite series to mathematical analysis. The many theorems in the subject are also quite beautiful in the austere aesthetic typical of mathematics.

[The sum of the geometric series $12 + 12(1/5) + 12(1/25) + 12(1/125) + \ldots$ is 15.]

SORTING AND RETRIEVING

Sorting and retrieving seem to be low-level bookkeeping skills requiring very little mathematical background. Although there is a kind of mindless pleasure in manually alphabetizing lists (something like simple knitting, I suppose) or in locating a reference in a large file cabinet, few people give much thought to the theoretical aspects of these activities. Determining the best way to perform them is, however, an interesting mathematical problem that has a good deal of practical importance.

Assume you've been asked to arrange a very large pile of numbered slips of paper in order. You could sequentially compare each slip to the ones you've already ordered, place it in its rightful position, and then do the same thing with the next slip from the pile. Alternatively, you could divide the pile into many very small stacks which you order by any method. Then you pair up these stacks and merge the orderings of each pair by comparing the first elements, the second elements, and so on. With these larger stacks, do the same thing—pair them and merge their orderings. In this way you steadily decrease the number, but increase the size of the ordered stacks until at the end you're left with one ordered stack and the task is complete.

Which way you sort these numbers may not matter much if you have a dozen or so entries, but if you have thousands or millions, there will be a huge difference in the time required. (I'm assuming that the sorter, a person or a computer, can do two things: compare two numbers and move a number from one spot to another.) The first method, which is called the insertion sort algorithm, requires in the worst case approximately N^2 steps (or time units) when there are N entries to sort, whereas the second, which is called the merge sort algorithm, requires about N $\times$ natural log(N) steps for the same number N of entries. If N is 100, N^2 is 10,000, whereas N $\times$ natural log(N) is only 460—already a substantial difference.

Retrieval algorithms designed to find and retrieve pieces of information from a list and then to relate them in various ways are often even more time-consuming than either of these two sort algorithms. (This is especially true when the items are quite similar, it being considerably easier to locate a needle in a haystack than in a great mound of other needles.) Some such algorithms may require 2^N steps to complete when there are N entries, and it is this fact in particular which brings us to the practical import of these notions. If we take N = 100 again, 2^N is approximately 1.3×10^{30}, a number of steps so huge that the algorithm is practically useless (as well as useless practically). It's not inconceivable that the failure of modern command economies is due as much to information-theoretic constraints as to political ones, the commissars finding it increasingly difficult to centrally coordinate exponentially burgeoning data on supplies, parts, logistics. (See also the entry on *complexity*.)

The problem is universal. Now that a laser printer can turn a personal computer into a publishing house or a type foundry, our ability to sort and retrieve information is falling ever further behind our ability to generate it. As business reports and research papers, news articles and standard periodicals, databases and electronic mail, textbooks and other books all increase rapidly, the number of interdependencies among them goes up exponentially. New ways to link, cross-reference, and prioritize them are necessary if we're not to drown in a sea of information.

We often have more information than we know how to think with. Computer scientist Jesse Shera's paraphrase of Coleridge is depressingly apt: "Data, data everywhere, but not a thought to think." More and more people rely solely on digests, reviews, abstracts, and statistics, but lack the conceptual tools to make sense of them. The most important sorting algorithm extant is a good, broad-based education.

STATISTICS—TWO THEOREMS

∞

In his book *Suicide,* the French sociologist Emile Durkheim demonstrated that the incidence of suicide in an area may be reasonably predicted on the basis of demographic data alone. Likewise the unemployment rate may be estimated on the basis of sampling (and various other economic indices). In fact, many sociological and economic forecasts are independent of psychological insights and principles and are based in large part on probabilistic grounds. Although events are seldom foreseeable in their particularity (who is going to commit suicide or who become unemployed), large ensembles of events are generally easy to describe beforehand. *Very* roughly, this is what two of the most important theoretical results in probability and statistics suggest. (See also the entries on *mean, correlation,* and *probability.*)

Less impressionistically, the law of large numbers states that the difference between the probability of some event and the relative frequency with which it occurs necessarily approaches zero. In the case of a fair coin, for example, the law, which was first described by Swiss mathematician Jakob Bernoulli in a work published posthumously in 1713, informs us that the difference between 1/2 and the quotient of the number of heads and the total number of flips can be proved to get arbitrarily close to zero as the number of flips increases without bound.

This shouldn't be interpreted to mean that the difference between the total number of heads and the total number of tails will get smaller as the number of flips increases; usually quite the opposite is the case. If one coin is flipped 1,000 times and another coin 1,000,000 times, the ratio of heads to flips will probably be much closer to 1/2 in the latter case despite the fact that the difference between the numbers of heads and tails will probably be much larger there as well. Fair coins comport themselves well in a ratio sense, but not in an absolute sense. And contrary to much barroom wisdom, the law of large numbers doesn't imply the gambler's fallacy: that a head is more likely after a string of tails. It's not.

Among other beliefs that the law does justify is the experimenter's trust that the average of a collection of measurements of some quantity will approach the true value of the quantity as the number of measurements increases. It also provides the grounds for the common-sense observation that if a die is rolled N times, the likelihood that the frequency of 5's obtained differs much from 1/6 gets smaller and smaller as N gets larger. Like the die, we are not predictable as individuals, but as a collective we're quite so. More broadly construed again, the law of large numbers provides theoretical support for the intuitive idea that probability is a guide to the world. The Nielsen TV ratings, the Gallup polls, insurance tables, countless sociological and economic studies all provide evidence of a probabilistic reality messier than that of coins and dice but no less genuine.

The other law I want to sketch here is called the central limit theorem, which states that the average or the sum of a large collection of measurements of any given characteristic is best described by a normal bell-shaped curve (sometimes called a Gaussian curve in honor of the great nineteenth-century mathematician Karl Friedrich Gauss). This is so even if the distribution of the individual measurements is not itself normal.

To illustrate, imagine a factory that produces disk drives for computers, and assume that the factory is run by a subversive hacker who ensures that about 30% of the drives break down after only 5 days, and the remaining 70% last for about 100 months before breaking down. The distribution of the life spans of these drives is clearly not described by a normal bell-shaped curve, but rather by a U-shaped curve consisting of two spikes, one at 5 days and a bigger one at 100 months.

Assume now that these drives come off the assembly line in random order and are packed in boxes of 36. If we took the time to calculate the average life span of the drives in a box, we'd find it to be about 70 months or so, maybe 70.7. Why? If we determine the average life span of the drives in another box of 36, we'd again find the average life span to be about 70 or so, perhaps 68.9. In fact, if we examine many such boxes, the average of the averages will be very close to 70, and what's more significant, the distribution of these averages will be approximately normal (bell-shaped), with the right percentage of boxes having averages between 68 and 70, or between 70 and 72, and so on.

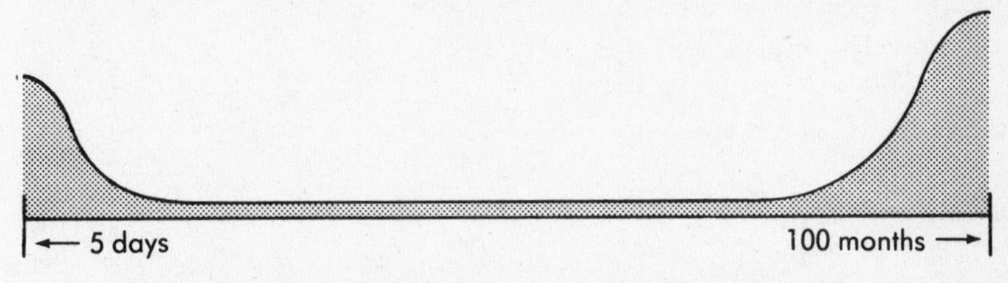

U-shaped distribution of disk drive lifetimes in a typical box of 36

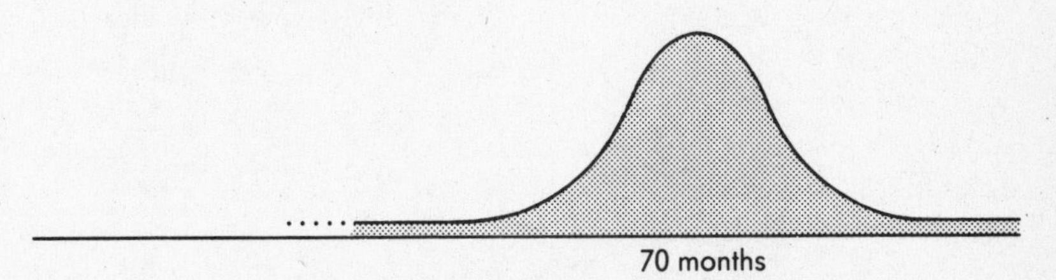

Normal bell-shaped distribution of the average lifetimes of the disk drives in many such boxes

The central limit theorem

The central limit theorem states that in a wide variety of circumstances this is to be expected—averages and sums of even nonnormally distributed quantities will themselves have a normal distribution.

The normal distribution also arises in the measuring process since the measurements of any quantity or characteristic tend to have a normal bell-shaped "error curve" centered on the true value of that quantity. Other quantities which tend to have a normal distribution might include age-specific heights and weights, natural gas consumption in a city for any given winter day, thicknesses of machine parts, IQs (whatever it is that they measure), the number of admissions to a large hospital on any given day, distances of darts from bull's-eyes, leaf sizes, nose sizes, or the number of raisins in boxes of breakfast cereal. All these quantities can be thought of as the average or sum of many factors (genetic, physical, or social), and thus the central limit theorem explains their normal distribution. To reiterate, averages or sums of quantities tend to have a normal distribution even when the quantities of which they're the average (or sum) don't.

SUBSTITUTABILITY AND MORE ON ROTE

∞

I know intelligent people who easily absorb the most complicated of legal arguments, the most nuanced of emotional interchanges, the most intricate of historical tales (even the most clumsy of relative clauses), and yet who, when they are confronted with a simple "word problem" in mathematics, immediately acquire an uncomprehending glaze over their eyes. They freeze and forget to ask the common-sense heuristic questions that they do in other realms of their life: Where did this problem come from? What do I really want to find and why? How can I simplify the situation or obtain an approximate answer? Is the problem related to something that I do know? Can I work backwards from the solution to the data?

Mathematical problems strike many of these people as requiring a mindless, rote mode of thinking and the instantaneous performance of some sort of calculation. If the answer doesn't immediately bite them on the nose, they assume they'll never find it. They would find it odd to engage in a mathematical discourse—to think about a mathematical problem in narrative terms. (See also the entry on *computation and rote.*) Like Molière's character who was shocked to find that he'd been speaking prose his whole life, many people are surprised when told that much of what they characterize as common sense or logic is mathematics.

These erroneous attitudes may derive in part from the fact that discourse in formal mathematics does have some peculiar properties that legal arguments, emotional interchanges, and historical tales do not. To illustrate one small but important example, let's look at the venerable practice going back to Euclid of substituting equals for equals.

If the perverse desire to do so ever overcomes us, we can always replace "25" with "5^2" or with "$(3^3 - 2)$" in a calculation and still get the same result. Likewise, if "equilateral triangles" were everywhere substituted for "equiangular triangles" in a mathematical discussion, the discussion would still make just as much sense, these expressions being different ways to refer to the same set of figures. In general, how we describe or denote mathematical objects does not at all affect the truth of the statements that contain these descriptions. Usually termed extensionality, this substitutability property may seem quite reasonable and obvious, but it is distinctive of formal mathematics.

We say that two sets of numbers are equal if they have the same members because how mathematical sets are described isn't important. By contrast, we say that two school clubs are different even if the same students happen to belong to both because how the clubs are characterized (their purpose) is crucial. Informal talk of beliefs, wants, purposes, and the like is said in logic to be intensional (with an "s"), not extensional, and does not allow substitutability. For example, if a geographically confused Easterner *believes* that Cheyenne is in Montana, then even though "Cheyenne" equals "the state capital of Wyoming," it certainly doesn't follow that he believes that the state capital of Wyoming is in Montana. The latter characterization of the city can't be substituted for the former in this intensional belief context.

The computer program which translates "The spirit is willing, but the flesh is weak" as "The vodka is agreeable, but the meat is too tender" or "three wise men" as "three wise guys" is ascribing an extensionality to natural languages (proverbs in particular) that they just don't possess. The family who want to arrive in Disney World by January 7 can't be said to want to arrive there by Millard Fillmore's birthday, even though it turns out that "January 7" and "Millard Fillmore's birthday" denote the same day. Again, substituting equals for equals fails to preserve the truth of the statement.

Let me stress, however, that the distinction between extensional and intensional contexts is not equivalent to that between mathematical and nonmathematical contexts. I was careful above to write that only formal mathematics is extensional; when manipulating symbols, checking derivations, or performing calculations, substitution is unobjectionable. But surely there is talk of wants, beliefs, and purposes in the interpretation and application of mathematics, and in these more human contexts mathematics too is intensional. Most mathematical study, whether by professionals or in everyday life, is spent learning how to prove the theorems of formal mathematics, how to interpret them in a specific situation, and how and when to apply the resulting rules and formulas. There are stories and values here—the source of the problem, its relation to other problems, its possible applications—and this is where the heuristics, the informal talk, the intensional contexts arise. In these respects mathematical thought is much more like that in law, history, literature, and daily living.

Every significant advance in mathematics has a story attached to it which gives meaning and significance to it—the Pythagorean theorem, the development of our number system, the Arab progress in algebra, the evolution of calculus from Isaac Newton to Leonhard Euler, non-Euclidean geometry, Galois theory, Cauchy's theorem in complex analysis, Cantor's set theory, Gödel's incompleteness theorem, and many, many other theorems and ideas. Are any of these merely computations or formal proofs? Computation and proof are usually taken as characteristic of mathematics, but as necessary as they sometimes are, most people most of the time do not want either. They want what they want in other fields: explanations, stories, heuristics.

The idea that mathematical thinking differs fundamentally from that in other fields is a pernicious one. One source for it lies with mathematics teachers who don't connect their subject with the rest of the curriculum, or with current events and news stories having a mathematical slant, or, indeed, with any aspect of the student's life. Another source derives from cultural misconceptions of the coldness, irrelevance, and difficulty of mathematics. And, as I've suggested here, still another oblique source is the philosophical misunderstanding that confuses the substitutability and empty extensionality of formal mathematics with the richer, intensional core of the subject (and exalts the former). Mathematics has just as much narrative, purpose, and storytelling in it as it has calculation and formulas. If we fail to see this and remain ignorant of mathematics but blindly reverent toward its techniques, we impoverish ourselves needlessly and empower others excessively.

SYMMETRY AND INVARIANCE

Symmetry and invariance are not so much topics in mathematics as guiding principles of mathematical aesthetics. From the Greeks' preoccupation with balance, harmony, and order to Einstein's insistence that the laws of physics should remain invariant for all observers, these ideas have vivified much of the best work in mathematics and mathematical physics.

Symmetry and invariance are complementary notions. Something is symmetrical to the extent that it is invariant under (or unchanged by) some sort of transformation. To illustrate, let's consider a circle. We can rotate it or reflect it about any of its diameters, and it still retains its circularity. Its symmetry consists in its invariance under these changes. But suppose now we squash it a bit (say by drawing the circle on a piece of soft wood and compressing the wood). We note that it assumes an elliptical shape. No longer are any two of its diagonals equal in magnitude; some diagonals are elongated, others shortened. This property of a circle is lost, but there are others that are not. For example, the center of the flattened figure still bisects each of its diameters, whatever their length. This latter property is invariant even under this more extreme sort of change and thus reflects a deeper sort of symmetry.

Observations of this kind suggested to the nineteenth-century German mathematician Felix Klein the idea that theorems about geometrical figures might be classified according to whether or not they remain true when the figures are subjected to various changes and transformations. More generally, given any specified collection of transformations (rigid motions in the plane, uniform compressions, projections), Klein asked what properties of figures remain invariant under these transformations. The body of theorems dealing with these properties is deemed the geometry associated with this collection of transformations.

Thus Euclidean geometry is thought of as the study of those properties left invariant by rigid motions: translations, rotations, and reflections. Projective geometry, on the other hand, is understood to be concerned with that smaller class of properties which is left invariant by all the rigid motions *plus* projections. (The projection of a figure is, roughly speaking, the shadow it casts when lit from behind. The projection of a circle might be an ellipse of some sort, for example.) And topology is the discipline devoted to the still smaller class of properties which are left invariant under the above transformations *plus* even the most extreme twistings and stretchings.

Length and angle are Euclidean properties (they are preserved by rigid motions), but they are not invariant under projective transformations. Linearity and triangularity are projective properties (they are preserved under projective transformations since lines and triangles are always transformed into other lines and triangles by projection), but they are not invariant under topological transformations. And connectedness and the number of holes in a figure are properties which persevere despite twistings and stretchings of all sorts.

This idea of deeper invariances marking more subtle symmetries is a very powerful one outside of geometry as well. Symmetrical art forms far more abstract than, say, the Greek key or the Alhambra in southern Spain are (in a Pickwickian sense at least) what modern art is about. Einstein's theory of special relativity (he considered calling it the theory of invariants) was a result of his determination that the laws of physics should be invariant under a group of transformations due to the Dutch physicist H. A. Lorentz.

One social consequence of mathematics' concern with timeless, enduring truths and this aesthetic of symmetry and invariance is that it is, in its purest form, necessarily aloof from the real world of quirky contingency and human idiosyncrasy. Even in applied or popular works on mathematics, this aversion to the personal is sometimes observable.

I remember receiving a letter from a mathematician saying that he very much enjoyed my books, but that they weren't mathematics since I freely used the word "I" in them. He had a point, of course, but the fact that he had to explicitly dissociate his enjoyment of the books from his devotion to pure mathematics is sad. Or so *I* feel.

The mathematician's strategy of searching for symmetry and invariance cannot fail since complete disorder at every level of analysis is a logical impossibility. Still, recognition of the asymmetrical, the changeable, and the personal can't hurt either.

TAUTOLOGIES AND TRUTH TABLES

∞

Either Aristotle had red hair or Aristotle did not have red hair. Since it's not true that Gottlob or Willard is present, both Gottlob and Willard are absent. Whenever Thoralf is out of town Leopold vomits, so if Leopold isn't vomiting, then Thoralf isn't out of town. Each of these astonishing insights is an instance of a mathematical tautology, a statement true by virtue of the meaning of the logical connecting words "not," "or," "and," and "if" . . . , then . . . " ("Tautology" is also used informally in a somewhat broader sense.)

If we symbolize by capital letters the simple declarative sentences of which these statements are composed, then they may be rendered as "A or not A," "If not (G or W), then (not G and not W)," and "If (if T, then L), then (if not L, then not T)." Any other simple declarative sentence may be substituted for A, G, W, T, or L and the result will still be a true statement. Thus, "Either Gorbachev is a transvestite or he isn't," "Because it's not true that George or Martha is guilty, then both George and Martha are innocent," and "Since whenever it rains the computer store is closed, if the computer store is open, then it's not raining" are all tautologically true statements. These particular tautologies even have names, being known as the law of the excluded middle, De Morgan's law, and the law of contraposition, respectively.

Logicians have formalized the checking process by which such statements are judged tautologies. They've developed rules, generally called truth tables, for each of the connectives: a "Not P" statement is true exactly when P is false; "P and Q" statements are true only if both P and Q are true; "P or Q" statements are true only if at least one of P or Q is; and "If P, then Q" statements are false only when P is true and Q is false. (It should be mentioned that there are other, nonmathematical uses of "if . . . , then . . ." which are interpreted differently. For mathematical purposes, however, it is convenient to assign truth to "If the moon were made of green cheese, then Bertrand Russell would be Pope" and, more generally, to assert that anything at all follows from a false statement.) And lastly, "P if and only if Q" statements are true only when P and Q have the same truth value, either both true or both false.

Two small digressions: What is the following advertisement for a faculty position in mathematics really saying? "We are searching for a candidate who is either an enthusiastic and effective teacher or one who is active in research. Unfortunately, we cannot consider those candidates who are enthusiastic and effective teachers who are not also active in research." Leaving academe now, imagine yourself on an island inhabited by people who either always tell the truth or always lie. You're at a fork in the road and you need to know which road leads to the capital. Luckily a local resident runs by (whether liar or truth-teller you don't know), and he has time to answer only one yes-or-no question. What question should you ask him to determine which road leads to the capital?

In propositional logic, the fragment of mathematical logic I'm describing, the opposite of a tautology is a contradiction, a statement which is always false by virtue of the meaning of its logical connectives. "Waldo is bald, and Waldo is not bald" may be judged untrue without knowledge of Waldo or his hairline. Formal contradictions such as "A and not A" and "(C and not B) and (if C, then B)" are false no matter which simple declarative sentences are substituted for A, B, and C. The third and largest category of statements in propositional logic comprises those which are sometimes true and sometimes false depending upon the truth or falsity of their constituents. Just to examine one example of a truth table, consider the table for "A and (B or not A)" on the following page.

The columns under A and B list the four possible assignments of truth and falsity to this pair of symbols. The first row is to be interpreted as saying that the statement as a whole, "A and (B or not A)," is true (indicated by the underlined T) whenever A and B are both true. The second through the fourth rows say that the statement is false for any other assignment of truth values to A and B. [The traditional symbols for "and," "or," and "not" are $\wedge$, $\vee$, and $\neg$, yielding $A \wedge (B \vee \neg A)$ for the above. An arrow, $\rightarrow$, is the symbol for "if . . . , then . . . ," while a two-sided arrow, $\leftrightarrow$, does service for "if and only if."]

A	B	A	and	(B	or	not	A)
T	T	T	T	T	T	F	T
T	F	T	F	F	F	F	T
F	T	F	F	T	T	T	F
F	F	F	F	F	T	T	F

Truth tables for determining truth and falsity are not really necessary for this statement (which turns out to have the same truth values as "A and B"), but they're often invaluable for more complicated assertions containing nested iterations (formulas within formulas) and many more than two-sentence symbols. Not only are such tables used to determine whether arbitrary statements are tautologies, contradictions, or contingent, they may also be employed to decide if certain types of argument made famous by Lewis Carroll are valid. (An example from the wonderland of economics: If the bond market rises or if interest rates decline, either the stock market drops or taxes aren't raised. The stock market drops if and only if the bond market rises and taxes are raised. If interest rates decline, then the stock market does not drop or the bond market doesn't rise. Therefore, either taxes are raised or the stock market drops and interest rates decline.) The procedures for checking this sort of validity are so routinized that circuits for "and," "or," and "not" are built into the hardware of all computers so that the machines almost instantaneously determine the truth of complex sentences and conditions.

The truth table mechanisms are of little use, however, for sentences containing relational phrases. (See the entry on *quantifiers.*) The statements "All friends of Mortimer are friends of mine," "Oscar is Mortimer's friend," and "Oscar is my friend," considered as statements of propositional logic, must be symbolized by single letters—say P, Q, and R, respectively. These symbols don't reflect the fact P and Q imply R since the implication does not depend on the meaning of "and," "or," "not," and "if . . . , then . . ." The implication is only captured in predicate logic, which encompasses not only propositional logic but also the logic of relational phrases ("is a friend of" in this case) and their associated quantifiers ("all" here). In this wider realm the American logician Alonzo Church has proved that there can never be a recipe (such as the truth table method) for determining the validity of sentences or arguments.

[The solutions to the puzzles: A little head thumping will convince you that the requirements for the position boil down to being active in research. Calculation shows the same thing, since whatever truth value we assign to E (enthusiastic, effective teaching) and R (active research) $(E \vee R) \wedge \neg(E \wedge \neg R)$ has the same truth value as R. And an appropriate question to pose to the truth-teller/liar is "Is it the case that the left road leads to the capital if and only if you are a truth-teller?" Both truth-teller and liar will answer yes if the left road leads to the capital and no if it doesn't. An alternate question you might ask: "If I were to ask you if the left road leads to the capital, would you say yes?" Again, both truth-teller and liar will give the same response.]

TIME, SPACE, AND IMMENSITY

Laurence Sterne's eighteenth-century novel *The Life and Opinions of Tristram Shandy, Gentleman* led Bertrand Russell to his "paradox of Tristram Shandy." The paradox concerns the narrator of the book, Tristram Shandy, who, Russell recalled, had taken two years to write the history of the first two days of his life. Shandy grieved that, at this rate, the latter parts of his life would never be recorded. Russell noted, however, that "if he had lived forever, and not wearied of his task, then, even if his life had continued as eventfully as it began, no part of his biography would have remained unwritten."

The resolution of the paradox depends on the peculiar properties of infinite numbers (see the entry on *infinite sets*). At the rate reported, the account of Shandy's third day would have taken him a year to write, as would that of his fourth, fifth, and sixth days. Each year he would have written a full account of another day in his life, and thus even though he would have fallen further and further behind each year, there is not a day which would go unrecorded provided he lived forever.

We may find it difficult to respond rationally to vastly differing time scales even if we stray only a finite distance from our contemporary cocoons. Attempting to reconcile astronomical, geological, biological, and historical time intervals can induce a feeling of frustration within nanoseconds, but it is nevertheless a worthwhile endeavor. Once allowance is made for scale, similar structures often appear. The sense of perspective which results may significantly influence our outlooks, if not our actions and decisions.

Similar perspective may be gained from spatial comparisons. Note that the earth's highest point, Mount Everest, is only about 6 miles high, a figure comparable in magnitude to the depth of the deepest oceans. Hence the earth's most extreme surface irregularities are less than 1/1,000 of its 8,000-mile diameter and correspond to bumps about 2/1,000 of an inch high on a 2-inch billiard ball (i.e., 1/1,000 × 2). Thus, despite its mountains, oceans, and irregular terrain, the earth is smoother (but not necessarily rounder) than the average billiard ball.

One of my favorite stories, Tolstoy's "How Much Earth Does a Man Need?," is not inappropriate here. About a man given the opportunity to possess all the land he can walk around in one day, the parable shows how the man's greed leads to his death and thus provides the answer to the question in the title. A man needs about 6 feet by 2 feet by 4 feet—enough for a grave. Even if we're considerably more generous, however, and provide every human being on earth with a cubical room 22.5 feet on a side, the volume of the Grand Canyon is sufficient to house all 5 billion of these cubes (see the entry on *areas and volumes*).

We naturally are most concerned with our familiar spatial and temporal surroundings, but that shouldn't blind us to their necessary parochialism. Recall that the nineteenth-century introduction of fast cameras resulted in pictures of people and animals in motion that were thought to be quite odd and counter-intuitive. We sometimes make too much of the relatively minor differences in the approach to time and planning adopted by American and Japanese businessmen or of the small variations in the temporal horizons of teenagers and senior citizens. We might occasionally ponder how we would relate to an alien or artificial being that, though much more intelligent than we, took 100,000 times as long to respond to stimuli. Communication with such entities would, at first glance, seem barely possible, yet something roughly like the inverse of this relation holds between slow, smart me and my fast, dumb computer. And the fastest computers are positively torpid compared with intra-atomic events; an electron circling a nucleus makes approximately 10^{15} orbits per second.

Of still vaster scale is the ancient Indian fable of a cubic stone one mile wide and one million times harder than diamond. Every million years a holy man comes by and gives the stone the tiniest caress. After a short while, the stone is worn away, a short while estimated to be 10^{35} years. The history of the universe, by contrast, is approximately 1.5×10^{10} years long.

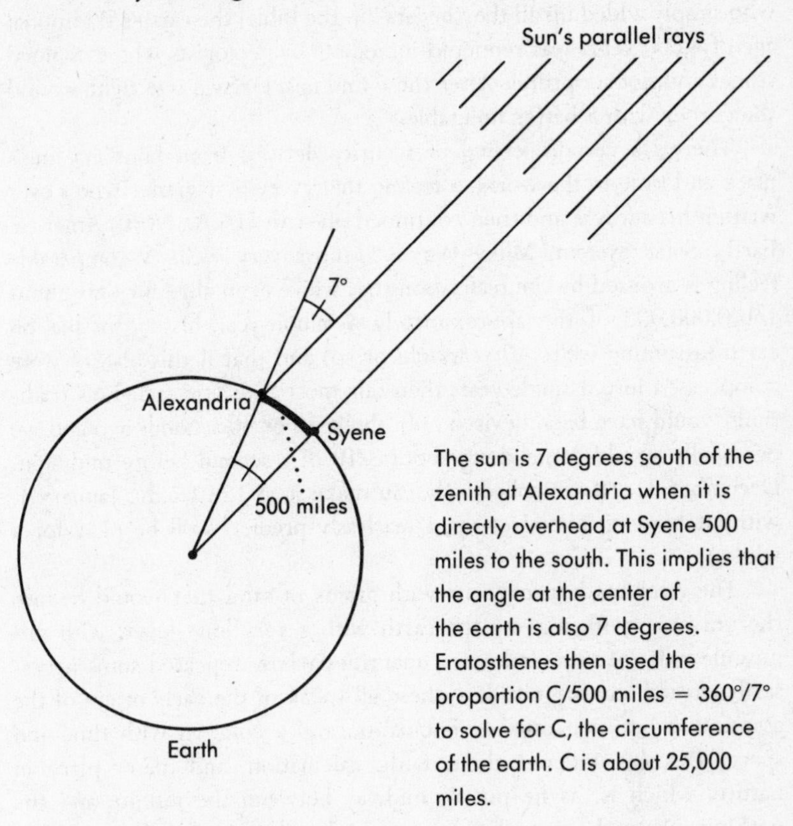

The sun is 7 degrees south of the zenith at Alexandria when it is directly overhead at Syene 500 miles to the south. This implies that the angle at the center of the earth is also 7 degrees. Eratosthenes then used the proportion C/500 miles = 360°/7° to solve for C, the circumference of the earth. C is about 25,000 miles.

Although it sometimes reminds me of the silly masculine practice of spending the first five or ten minutes at a party discussing one's route to the party, I've always enjoyed studying time lines and spatial comparisons that encapsulate and relate the various orders (astronomical, geological, biological, geographical, historical, and microphysical). Simplistic at times, these charts are nevertheless quite useful in helping us to orient ourselves to the cosmos. Eratosthenes' calculation in 200 B.C. of the circumference of the earth was remarkable in this respect. He deduced the value from the fact that the sun is 7 degrees south of the zenith at Alexandria when it is directly overhead at Syene 500 miles to the south. One of the great pillars of our present worldview, the theory of evolution, developed out of the increasing untenability of the biblical time frame brought about by geological research. Obtained by scholars who simply added up all the "begats" in the Bible, the earth's traditional age of 4,000 years was rendered incredible by geologists who examined stones and not scriptures. After these findings, Darwin was right around the corner with a better timetable.

There's a certain feeling of security derived from knowing one's place and time in the world, a feeling that every first-grader who's ever written his address and then continued on with U.S.A., North America, Earth, Solar System, Milky Way . . . knows very well. A comparable feeling is aroused by the realization that we've been alive for only about 1/100,000,000 of the approximately 4-billion-year history of life on earth (assuming we're 40 years old or so) and that if this history were compressed into a single year, then our most "ancient" religious traditions would have been devised only about 30 or 40 seconds ago and we personally would come along about 3/10 of a second before midnight, December 31. (If we collectively can make it to 12:01 a.m., January 1, without blowing ourselves up, I fearlessly predict we'll be okay for a while.)

The Archimedean concern with grains of sand that would fit into the universe; with moving the earth with a very long lever; with minuscule units of time and other quantities whose repeated sums necessarily exceed any magnitude—these all speak of the early origin of the association between number fascination and a concern with time and space. Pascal wondered about faith, calculation, and man's place in nature, which is, as he put it, midway between the infinite and the nothing. Nietzsche speculated about a closed and infinitely recurring universe. Henri Poincaré and others with an intuitionist or constructivist approach to mathematics have compared the sequence of whole numbers to our pre-theoretic conception of time as a sequence of discrete instants. From Riemann and Gauss to Einstein and Gödel, mathematicians have made conjectures about space and time. These topics have in fact been a staple of mathematico-physical reflection for millennia.

There is no conclusion from this inchoate discourse except for the perhaps feeble one that such deliberations are somehow "good" for us —somehow therapeutic, grounding, or sobering. Speaking of "sobering," I'm annoyed by people who after such a discussion and a few drinks either retreat behind some dogma (not always religious) or else get maudlin and mumble something like "What difference does it all make? What will it matter in 50,000 years?" One might reasonably react with stoicism and resignation to this fatalistic question. But consider this. Maybe nothing we do now will matter in 50,000 years, but *if* that is so, then it would seem that nothing that will be the case in 50,000 years makes a difference now either. In particular it doesn't make a difference now that in 50,000 years what we do now won't make a difference.

TOPOLOGY

∞

According to Woody Allen, fake rubber inkblots were originally 11 feet in diameter and fooled nobody. Later, however, a Swiss physicist "proved that an object of a particular size could be reduced in size simply by 'making it smaller,' a discovery that revolutionized the fake inkblot business." This little tale could be interpreted as a parody of topology, a subject whose insights at first look do seem a little obvious. It is, after all, a branch of geometry concerned only with those basic properties of geometric figures that remain unchanged when the figures are twisted and distorted, stretched and shrunk, subjected to any "schmooshing" at all as long as they're not ripped or torn.

Rather than give a technical definition of "schmooshing," I'll proceed with some observations and examples. Size is not a topological property since, as Allen's physicist observed, spheres (or rubber inkblots) can be contracted or expanded into smaller or larger ones without ripping (simply by making them smaller or larger). Think of blowing up or removing air from a balloon. Neither is shape a topological property, since a spherical balloon (or odd-shaped rubber inkblot) can be squeezed into an ellipsoid or a cube or even a rabbit shape without tearing it.

Since the properties of a rubber sheet that persist after it's been pulled, pushed, and deformed are topological ones, topology has been called "rubber sheet" geometry. (The phrase is linked in my mind with my high school calculus teacher in Milwaukee, who attended a summer program on "modern mathematics" and thereafter ascribed any difficulties encountered by his students to an ignorance of rubber sheet geometry. He was always stretching a large rubber band as if this somehow illustrated the incontrovertibility of his contention.)

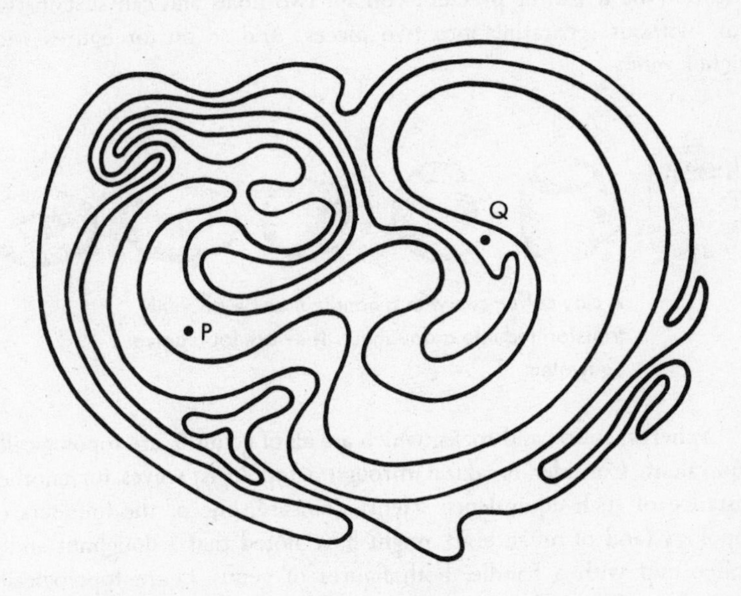

A closed curve on a plane divides the plane into two parts, the inside and the outside. Is P on the inside or the outside? Q?

Whether a closed curve in space, say an unbroken piece of thread, has a knot in it or not is a topological property of the curve in space. That a closed curve lying on a flat plane, no matter how convoluted the curve, divides the plane into two parts—the inside and the outside—is a topological property of the curve in the plane. The dimensionality of a geometric figure (how many dimensions it possesses), whether or not it has a boundary and if so of what sort—these too are topological properties. (See the entry on *Möbius strips.*)

Also a matter of some significance is the genus of a figure: the number of holes it contains or, from a more butcherly perspective, the maximum number of cuts passing all the way through the figure it can sustain without separating into two pieces. A sphere has genus 0 since it contains no holes and even one cut will separate it into two pieces. A torus (a doughnut or tire-shaped figure) has genus 1 since it contains one hole (the hole in the doughnut) and can sustain one cut without separating into two pieces. Figures of genus 2, such as eyeglass frames without the lenses or pretzels, contain two holes and can sustain two cuts without separating into two pieces. And so on for figures with higher genus.

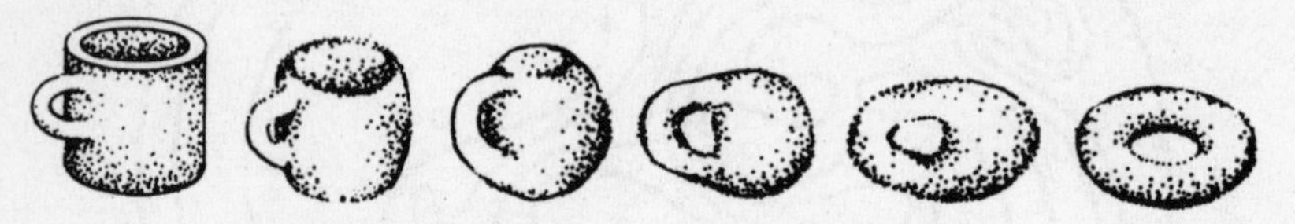

A clay coffee cup with a handle may be smoothly transformed into a doughnut. They are topologically equivalent.

Spheres, cubes, and rocks, which are all of genus 0, are topologically equivalent. Consider breakfast through a topologist's eyes for another instance of such equivalence. Henri Poincaré, one of the founders of topology (and of much else), might have noted that a doughnut and a coffee cup with a handle, both figures of genus 1, are topologically equivalent. To see this, imagine a coffee cup made of clay. Flatten the body of the cup and expand the size of its handle by squeezing clay from the body into the handle. The finger hole of the cup's handle is in this way transformed into the hole of the doughnut, and we can then easily recognize the topological equivalence. Human beings are, at least on a gross level, also of genus 1. We're topologically equivalent to doughnuts, our alimentary/excretory canal corresponding to the hole in the doughnut. (This latter fact has less appeal, however, as idle breakfast speculation.)

There are applications of these ideas, but they are for the most part internal to mathematics itself. Often in theoretical work, for example, what is important is knowing that a solution exists and not necessarily having a method for finding it. To get a flavor of such so-called existence proofs, imagine that a mountain climber begins his ascent at 6 a.m. Monday and arrives at the summit at noon. Tuesday morning he starts down at 6 o'clock and reaches the base at noon. We make no other assumptions about how fast or how smoothly the climber travels on the two days in question. He might, for all we know, have climbed at a slow pace and rested often on his way up on Monday, and after a leisurely stroll around the summit Tuesday morning descended at a literally breakneck pace, managing to fall the last 1,000 feet. The question now is: Can we be certain that no matter how he climbs there will necessarily exist some instant between 6 a.m. and noon on the two days when the climber is at exactly the same elevation?

The answer is yes, and the proof is vivid and convincing. Imagine the ascent and descent, exact in every detail, being made simultaneously by two climbers. One climber starts at the base and the other at the summit and they both begin their journeys at 6 a.m. of the same day, mimicking what the original climber did on Monday and Tuesday, respectively. It's clear that these two climbers will pass each other going in opposite directions and at that instant their elevations will be the same. Since they're only reenacting the original climber's ascent and descent, we can be certain that the original climber was at the same elevation at the same time on the two successive days.

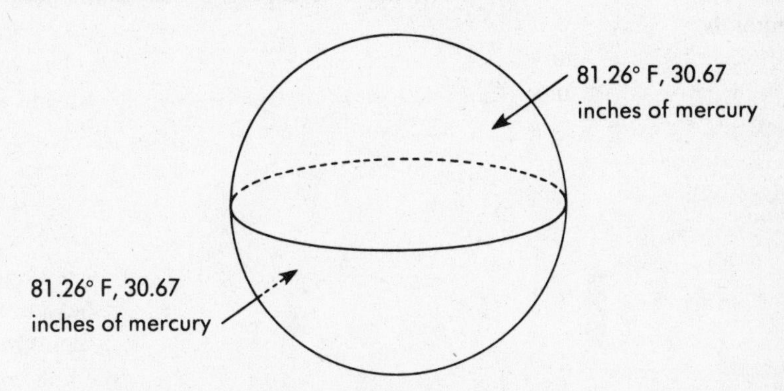

There is always a pair of antipodal points having exactly the same temperature and pressure.

A more counter-intuitive example of an existence theorem is the result that there are at all times points on opposite (antipodal) sides of the earth that have the same temperature and the same barometric pressure. These points move about and we have no way to find them, but it can be proved that they always exist. This is not a meteorological phenomenon but a mathematical one. Another example: Take a rectangular piece of tissue paper and place it flat on the bottom of a box, taking care to ensure that every inch of the bottom is covered. If you now crumple the tissue paper into a little ball but leave it in the box, you can be topologically certain that at least one point of the paper is directly above the same point on the bottom of the box that it was above before the paper was crumbled. The existence of such a fixed point is guaranteed.

Theorems such as these do sometimes lead to concrete and practical developments in fields such as graph and network theory, which are concerned with, among other things, mathematical idealizations of street and highway networks. As mentioned, however, they contribute more frequently to theoretical advances in other areas of mathematics. Algebraic topology, for example, uses topological and algebraic ideas to help characterize various geometric structures, while differential topology uses techniques from differential equations and topology to study very general sorts of higher-dimensional manifolds (surfaces). And catastrophe theory, a subdiscipline of differential topology, concerns itself with the description and classification of discontinuities—jumps, breaks, switches. There is much more to topology than fake rubber inkblots.

TRIGONOMETRY

Trigonometry dates from the Greeks and was used by Eratosthenes to calculate the circumference of the earth, by Aristarchus to estimate the distances of the moon and the sun, and by many others to dig straight tunnels and determine the distance of inaccessible points. For centuries astrologers used trigonometry to construct their charts and calculations, and even today many exploit the esoteric image of celestrial trigonometry to disguise the baselessness of their astrological beliefs. The elementary part of the subject, which will be our primary focus here, deals with right triangles and the ratios of their various sides, topics which might not at first glance, or even second glance, seem all that engaging. Still, it's on the syllabus (usually in too prominent a role, I think), so let's persevere and consider a canonical example. Assume you're 80 feet from the base of a TV tower whose height you need to know. Assume further that the angle of elevation of the tower, the angle your eyes make with the ground when you're looking at its top, is 45 degrees. How high is the tower?

The basic insight of elementary trigonometry is that similar triangles, triangles having proportional sides, are such that if you determine the ratio of one side to another in one of these triangles, you will find it to be equal to the ratio of the corresponding sides in any of the other similarly shaped ones. Thus, in right triangles whose other angles are both 45 degrees (remember, the sum of the angles of a triangle is always

180 degrees and similar triangles always have equal angles), the ratio of one of the shorter sides to the other is always 1 to 1 no matter how big or small the triangles are. Since you're 80 feet from the tower, you conclude that the TV tower is 80 feet high.

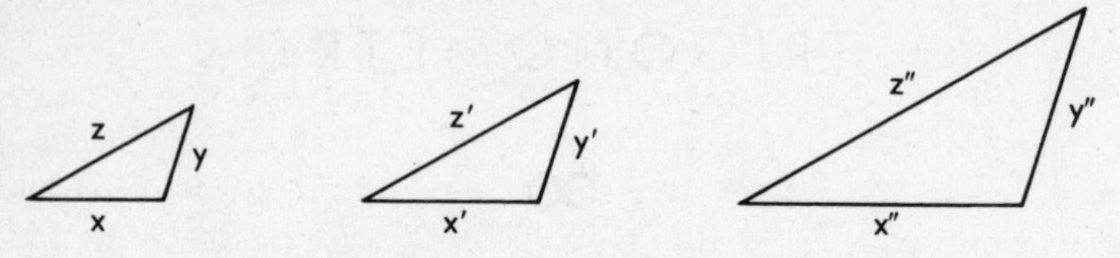

In similar triangles corresponding sides are proportional.

$$\frac{x}{y} = \frac{x'}{y'} = \frac{x''}{y''};\ \frac{y}{z} = \frac{y'}{z'} = \frac{y''}{z''};\ \frac{z}{x} = \frac{z'}{x'} = \frac{z''}{x''}$$

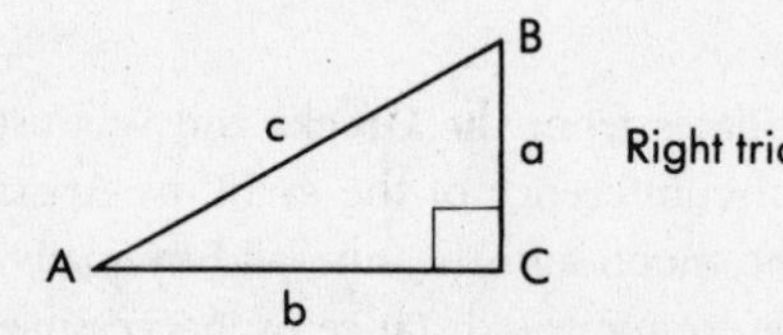

Right triangle ABC

The tangent of angle A is $\frac{a}{b}$; $\tan(A) = \frac{a}{b}$.

The sine of angle A is $\frac{a}{c}$; $\sin(A) = \frac{a}{c}$.

The cosine of angle A is $\frac{b}{c}$; $\cos(A) = \frac{b}{c}$.

The most common trigonometric functions

More generally, the tangent of an acute angle (less than 90 degrees) in a right triangle is defined to be the ratio of the side opposite the angle to the side adjacent to it. In the TV tower example the tangent of the angle of elevation is the ratio of the height of the TV tower to your distance from it, and when this angle is 45 degrees, the tangent is 1 to 1, which, expressed as a quotient, is 1/1 or simply 1. If you were 80 feet from the tower, but its angle of elevation was only 20 degrees, then you could consult a trigonometric table to determine that the tangent of this angle is approximately .36 to 1, or just .36. In this case the TV tower would be only .36 × 80 or about 29 feet tall.

The tangent of an angle is just one of a number of trigonometric functions with somewhat soporific names (sine, cosine, and so on) that associate ratios and numbers with angles. (I once knew someone who referred to the sine of X as Sominex.) Frequently we know some angles and some lengths and using these functions we can solve for the unknown angles or lengths we wish to find. The sine of an acute angle in a right triangle is defined to be the ratio of the side opposite the angle to the hypotenuse (the side opposite the right angle), whereas the cosine of an angle is the ratio of the side adjacent to the angle to the hypotenuse. In the case of a 45-degree angle the sine and cosine are equal, both approximately .71, while for a 20-degree angle the sine is .34 and the cosine is .94. Symbolically we write: $\sin(20^\circ) = .34$ and $\cos(20^\circ) = .94$, while $\tan(20^\circ) = .36$ and $\tan(45^\circ) = 1$.

The values for the trigonometric functions of any angle can be found by consulting tables that have been compiled for that purpose, by using infinite series formulas (see the entry on *series*), or, most easily, by pushing the appropriate buttons on a calculator. The ancient and medieval world did not have tables, formulas, or calculators and had to resort to complicated geometrical methods to find the sine, cosine, and tangent of the angles they were interested in, but their reasoning wasn't too different from that used in our TV tower example or in present-day surveying, navigation, and astronomy.

I should interject here that despite much freshman thought to the contrary, the sine of a 60-degree angle is neither twice the sine of a 30-degree angle nor three times the sine of a 20-degree angle. Ditto for cosines and tangents. Trigonometric functions do not grow in this linearly proportional manner. Furthermore, since they are functions and not products, the cancellation of the 20° in $\sin(20^\circ)/20^\circ = \sin$ is indeed a mathematical sin, but makes no mathematical sense. (See the entry on *functions.*)

As often happens, a discipline's focus shifts and definitions are generalized. In the case of trigonometry it was necessary to expand the definitions of the trigonometric functions to deal with triangles which were not right triangles and to cover the case of obtuse angles larger than 90 degrees. These more encompassing definitions are stated in terms of circles and rotation and link elementary trigonometry with its more modern incarnations.

The sine of an angle varies. It is 0 for an angle of 0 degrees, grows steadily but not linearly to a maximum for 1 for an angle of 90 degrees, shrinks back to 0 at 180 degrees, becomes more and more negative, sinking to a value of −1 at 270 degrees, gradually returns to 0 at 360 degrees, and begins this cycle again for angles greater than 360 degrees. (An angle of 370 degrees is indistinguishable from one of 10 degrees.) This periodic variation of the sine of an angle leads, when graphed, to the familiar sinusoidal form or sine wave, which describes many physical, particularly electrical, phenomena.

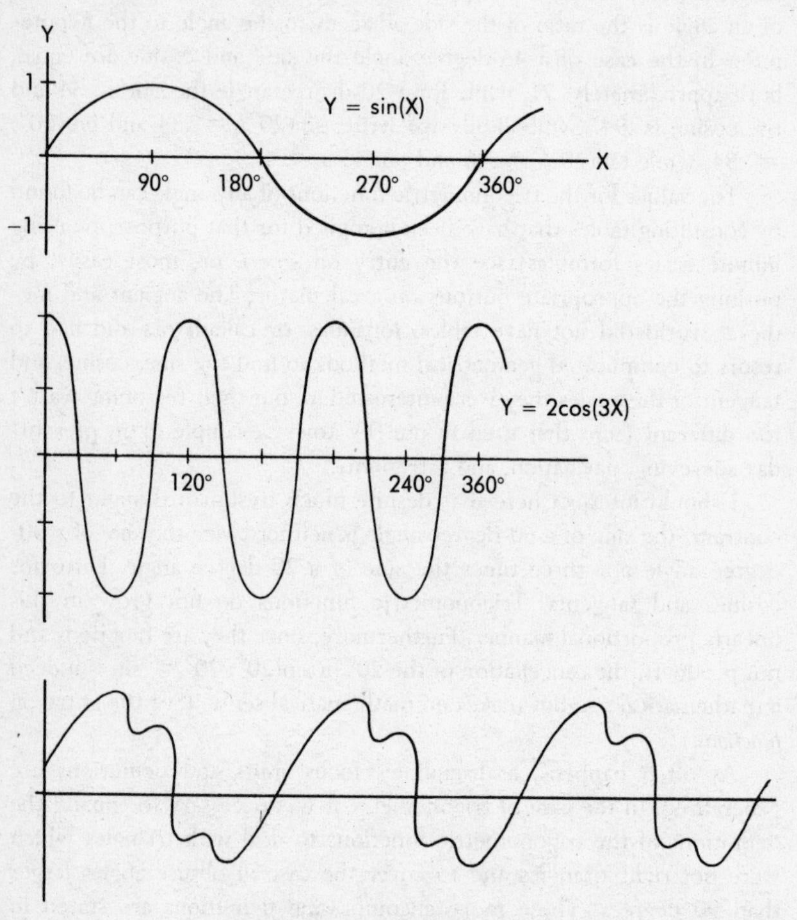

Sum of several different sine waves

Modern trigonometry is more concerned with periodicity and other properties of the trigonometric functions than it is with their interpretation as ratios. In his pioneering study of heat the French mathematician Joseph Fourier (moronic math joke: the name is pronounced yea yea yea yea just as tran tran tran tran is a well-known computer language) summed sines and cosines of various frequencies (degrees of wiggliness) to mimic the behavior of periodic, but nontrigonometric functions. By way of analogy imagine combining the sounds of two tuning forks vibrating at different frequencies in order to attain a sound which is the "sum" of the two sounds. Such summing techniques enable us to approximate a large class of important functions, even nonperiodic ones, with infinite Fourier series of trigonometric functions, and almost two centuries of work on such Fourier series have resulted in an indispensable computational tool for science and engineering, as well as in deep and subtle theorems in pure mathematics.

Still, many people, if they have any memory of the subject at all, remember trigonometric identities indicating the unconditional equality of one complicated trigonometric expression for another. Often an inordinate amount of time is spent in trigonometry class mastering the formal manipulations needed to demonstrate these equalities. Only a few of the identities are crucial, however, the most important one probably being the statement that the square of the sine of any angle added to the square of the cosine of that same angle always equals 1. Checking this with the values mentioned earlier, we see that for a 45-degree angle $.71^2 + .71^2 = 1$, and for a 20-degree angle $.34^2 + .94^2 = 1$.

I end with a little brain teaser and a story. The puzzle has grown less urgent with the spread of digital watches. You have a wristwatch whose hands move smoothly and continuously, and you note that it is 3 o'clock. How long will it be before the minute hand catches up to the hour hand? The approximate answer, it's clear, is between 15 and 20 minutes, but what is the precise answer? Now for the story: Recently I was listening to a psychologist on talk radio. His guest, a minor celebrity, went on at mind-numbing length about how he had hit bottom with drugs and alcohol but, through spiritual growth and a stint at a rehabilitation clinic, had recently "managed to turn my life around 360 degrees."

[The solution to the puzzle: Consider the angle, call it X, that the hour hand has moved through when the minute hand overtakes it. Since during this time the minute hand moves 12 times as far (it makes a complete revolution while the hour hand only moves between 3 and 4, 1/12 of the way around the face), the angle the minute hand has moved through is 12X. But another way to characterize the angle the minute hand has moved through is (X + 90°), the angle X the hour hand has moved through plus 90°, reflecting the 90° head start that the hour hand enjoyed at 3 o'clock. Setting these two different ways of describing the angle the minute hand has moved through equal, we obtain $12X = (X + 90°)$. Solving, we get $X = 8.1818°$. Translating degrees into minutes and seconds (90° = 15 minutes), we find that the minute hand overtakes the hour hand at 3:16:22 o'clock.]

TURING'S TEST, EXPERT SYSTEMS

The British mathematician and logician Alan M. Turing was the author of a number of seminal papers on logic and theoretical computer science. In one paper written in 1936 before any programmable computer had ever been built, he symbolically described the logical structure any such machine would have to possess. His description of an idealized computer specified in mathematical terms the relations among the input, output, actions, and states of what has come to be called a Turing machine. In another paper he argued that whether the physical substrate of such a machine is composed of neurons, silicon chips, or Tinkertoys is unimportant. Pattern and structure are what's crucial, not what the underlying stuff is. Alternatively put, the material is immaterial.

During World War II, Turing worked on cryptography and was instrumental in cracking the German code. He returned to abstract work afterward and in a well-known 1950 paper he proposed that the vague question of whether computers could be considered conscious be replaced with the less metaphysical one of whether a computer could be programmed to "fool" a person into believing he was dealing with another human being. Via a TV monitor someone would pose yes-or-no or multiple-choice questions to both an appropriately programmed computer and another person. He would then try to determine which set of answers came from the computer and which from the human. If he was unable to do so, the computer would be said to have passed the Turing test. Often the test is framed in terms of conversations. Picture yourself carrying on a conversation via monitor with two interlocutors. It would be your job to decide which one's hardware (or physiology) is based on silicon and which one's on carbon. (In this last regard, a suggestive analogy between the hardware-software distinction in computer science and the brain-mind distinction in philosophy has been developed by American philosopher Hilary Putnam.)

However phrased, the criteria for passing the Turing test are considerably clearer than are those for machine consciousness. Nevertheless, despite Turing's prediction that a computer would pass his test by the year 2000, none has even come close. Certainly for the foreseeable future, a computer's "conversation" will quickly reveal its mechanical soul. The amount of tacit knowledge that we all possess overwhelms our would-be imitators. We know that cats don't grow on trees, that one doesn't put mustard in one's shoes, that toothbrushes are not nine feet tall and sold in hardware stores, that even though fur coats are made out of fur and cloth coats out of cloth, raincoats are not made out of rain. All that would be necessary to expose the impostor machine would be to ask it about a few of the zillions (i.e., indefinitely many) of such humanly obvious understandings.

To achieve a different angle on the enormity of the programmer's task, imagine that in the course of his conversation with his two interlocutors our volunteer makes reference to a man's touching his head. How is a computer to evaluate the possible meaning of this gesture? Touching one's hand to one's head may mean that the person has a headache; or that the person is a baseball coach giving a signal to the batter; or that the person is trying to hide his anxiety by appearing nonchalant; or that the man is worried about his hairpiece slipping; or indefinitely many other things, depending on indefinitely many ever-changing human contexts.

There are, of course, so-called expert systems, special-purpose programs that do everything from analyzing large molecules to making certain sorts of medical diagnoses, from constructing legally binding documents (I have one that draws up wills) to performing complicated statistical analyses, from keeping track of mammoth databases to playing chess. A classic program (if the phrase means anything in such a young field) called ELIZA even mimics the evasive parrying of a nondirective therapist and is amusingly accurate for a minute or two.

These expert systems are written by "knowledge engineers" (a frightening term if there ever was one). Skilled in the techniques of artificial intelligence (programming designed to induce responses which, were they given by a human, would be deemed intelligent), these programmers interview experts in a given field, say petroleum geology. They try to capture some of the geologists' expertise in a form suitable for a computer: long lists of statements about rocks of the form "If A, B, or C, then D; otherwise E unless F," and complicated networks of interrelated geological facts. Later, if all goes well, the expert system program will be able to answer questions about where to dig for oil.

It is all the more remarkable, then, that with all the impressively recondite things that computers routinely do, it is conversation, mundane garden-variety chitchat, joking, and banter, that has proved so resistant to computer simulation. (See the entries on *substitutability* and *complexity*.) Simulating intercontinental trajectories is easy compared with simulating intrafamilial gossip. The latter requires an incomparably more flexible general-purpose program. Having eavesdropped on too many conversations whose participants would be hard pressed themselves to pass the Turing test, I should perhaps temper my human chauvinism a bit. Still, I find it heartening that many workers in the field of AI (as artificial intelligence is often referred to), having faced the task of imparting a semblance of general intelligence to machines, seem more respectful and cognizant of human complexity than do some literary theorists. They have had to take full account of a program's intentions and purposes in a way that contrasts starkly with the efforts of deconstructionists, for example, to eliminate both from their reductionist formal analyses of literary texts.

Whether AI will move beyond special-purpose expert systems and fulfill its promise (threat?) or ultimately be seen as something of a vast intellectual con game won't be clear for a long while. If true AI is attained, however, we should marvel at how lifelike these machines have become, not at how mechanical we've always been. We should think of ourselves as human Pygmalions bringing computer Galateas to life, not as automata whose mechanistic basis has been exposed by our computer progeny.

VARIABLES AND PRONOUNS

∞

A variable is a quantity that can take on different values but whose value in a given situation is often unknown. It is to be contrasted with a constant quantity. The number of biological parents a person has is a constant. The number of biological offspring he or she has is a variable.

Surprisingly, it wasn't until the late sixteenth century that the French mathematician François Viète came upon the retrospectively obvious idea of using letters to stand for variables (usually X, Y, and Z for real numbers, N for whole numbers). Though generations of beginning algebra students have grumbled at the introduction of variables, their use isn't any more abstract than the use of pronouns, to which they bear a strong conceptual resemblance. (Nouns, by contrast, are the analogues of constants.) And just as pronouns make communication easier and more flexible, variables allow much greater generality than does restricting our mathematical discourse to constants.

Consider the following sentence. "Someone once gave his wife something which she found so distasteful that she dropped it into the nearest garbage can and never willingly mentioned it to him again although he occasionally asked her about its whereabouts." Without pronouns the sentence would be unwieldy: "This person once gave this selfsame person's wife this thing and this person's wife found this thing so distasteful that this person's wife dropped this thing into the nearest garbage can and never willingly mentioned this thing to this person again although this person occasionally asked this person's wife about this thing's whereabouts." The introduction of variables restores some manageability to the sentence: "X gave X's wife Y a Z and Y found Z so distasteful that Y dropped Z into the nearest garbage can and never willingly mentioned Z to X again although X occasionally asked Y about Z's whereabouts."

A briefer example is provided by the injunction to Oscar: "Help whoever helps you." Without pronouns it would be replaced by "Help George if George helps Oscar, help Waldo if Waldo helps Oscar, help Martha if Martha helps Oscar, help Myrtle if Myrtle helps Oscar, and so on."

Few people have difficulty with pronouns or their referents, so it would seem that few should have difficulty with variables. In mathematics, however, constraints are placed on the variables which often enable us to determine their value. If $X - 2Y + 2(1 + 3X) = 31$ and $Y = 3$, we can find X. The techniques used to solve these and other, more complicated equations are what often prove puzzling. There is nothing obviously analogous in our everyday discourse with pronouns, although mystery whodunits aren't that dissimilar. Consider the following: Whoever (some Mr. X or Ms. X) canceled the guests' hotel reservations knew that they were coming for the celebration, that they'd be arriving late, and that not having a reservation in their name would be annoying to them and embarrassing to their hosts. If we know the principals involved, can we discover who canceled the reservations (that is, who equals X)? I maintain that the techniques and approaches employed in clarifying or solving these and other little human dramas are at least as complex as are those used in mathematics. (See also the entry on *substitutability*.)

One last editorial comment: Some have argued that the theoretical nature of mathematics distances us from our humanity and is somehow inconsistent with the spirit of compassion. As I've suggested in this entry and elsewhere, however, our use of language contains all the abstraction and complexity of mathematics. The "problem" with mathematics is not that it is abstract; it is that its abstraction is too often ungrounded, without human rationale. In matters of social policy or personal decision making, mathematics can help determine the consequences of our assumptions and values, but it is we (we X's), not some mathematical divinities, who are the origin of these assumptions and values.

VOTING SYSTEMS

How do democratic societies make decisions? The answer is "by voting," but what does this mean especially if, as is usually the case, there are more than two possible choices? Since one good illustrative example is often worth pages of careful exposition, assume, for illustration's sake, that there are five candidates for president of a small organization. Each of the group's members rank the five candidates, but the winner, as we shall see, may depend critically on which voting procedure is used.

It's necessary to be numerically specific here, so let's accept that there are 55 voting members and that their preferences (first, second, and so on) are as follows:

18 members prefer A to D to E to C to B
12 members prefer B to E to D to C to A
10 members prefer C to B to E to D to A
9 members prefer D to C to E to B to A
4 members prefer E to B to D to C to A
2 members prefer E to C to D to B to A

Supporters of candidate A might argue that the plurality method, whereby the candidate with the most first-place votes wins, should be used. With this method, A wins easily.

Supporters of B might argue that instead there should be a runoff between the two candidates receiving the most first-place votes. Candidate B handily beats A in such a runoff (18 members preferring A to B, but 37 preferring B to A).

Candidate C's people must think a little longer to come up with a method under which C will come out on top. They suggest that we first eliminate the candidate with the fewest first-place votes (E in this case), and then adjust the first-place preferences for the others (still 18 for A, now 16 for B, now 12 for C, still 9 for D). Next we eliminate the candidate among the four remaining having the fewest first-place votes (D in this case) and again adjust the first-place preferences for the remaining candidates. (C now has 21 first-place votes.) We continue this procedure of winnowing the candidates by removing at each stage the one with the fewest first-place votes. Using this method, C is elected.

Now the campaign manager of candidate D remonstrates that more attention should be paid to the overall rankings, not just to the top preferences. He argues that if first-place votes are each accorded 5 points, second-place votes 4 points, third-place 3 points, fourth-place 2 points, and last-place 1 point, then associated with each candidate will be a number, the so-called Borda count, which will accurately reflect that candidate's support. Since D's Borda count of 191 is higher than anyone else's, using this method he wins.

Candidate E, being a more macho sort, responds that only man-to-man (or man-to-woman) contests should count, and that pitted against any of the other four candidates in a two-person race, he comes out the winner. He claims that he therefore deserves to be the overall winner. (Someone like E who beats every other candidate in this way is called the Condorcet winner. Frequently, however, the voting is so muddled that no candidate emerges as the Condorcet winner.)

Who should be declared the winner, and what should be the group's preference ranking of the five candidates? The members might try to resolve the above impasse by voting on which method is to be used, but what method would they use in deciding this question? It's at least conceivable that the same problem would reappear at this higher level since the proponents of the various candidates might vote for that method which would result in their person being declared the winner.

(This natural tendency to tailor one's approach to one's self-interest reminds me of the advice the old lawyer gave to his protégé. "When the law's on your side, pound the law. When the facts are on your side, pound the facts. And when neither is on your side, pound the table." I should also note that the problem of deciding who votes is even thornier than that of deciding upon a voting system. In general, people want the law to enfranchise as many supporters and disenfranchise—or at least discourage—as many opponents as is reasonably possible. Opposition to women's suffrage and the practice of apartheid are examples of the latter, while the time-honored custom of stuffing the ballot box illustrates the former. Not limited to sleazy municipal election frauds, variants of the custom can tempt even the most high-minded regardless of their political orientation. Anti-abortion activists sometimes enlist the "votes" of the unborn, while environmentalists often go further and appeal to the "electoral" support of unconceived future generations.)

The situation with regard to voting systems isn't always as confused as the above example suggests. The numbers in the example (due to William F. Lucas via the eighteenth-century philosophers Jean-Charles de Borda and the Marquis de Condorcet and later theorists as well) were concocted in order to demonstrate the way in which the voting method used can sometimes determine the winner. But even though such anomalies may not always result, every method of voting is subject to them.

In fact, the mathematical economist Kenneth J. Arrow has demonstrated that there is never a foolproof way to derive group preferences from individual preferences that can be absolutely guaranteed to satisfy these four minimal conditions: if the group prefers X to Y and Y to Z, then it prefers X to Z; the preferences (both individual and group) must be restricted to available alternatives; if every individual prefers X to Y, then the group does too; and no individual's preferences dictatorially determine the group preferences.

Although every voting method has undesirable consequences and fault lines, some systems are better than others. One that might be especially appropriate for presidential primaries in which there are many candidates running is called approval voting. Under this system, each voter can vote for, or approve of, as many candidates as he or she wants. The principle of "one person, one vote" is replaced with "one candidate, one vote," and the candidate receiving the highest approval rating is declared the winner. Scenarios in which, for example, two liberal candidates split the liberal vote and allow a conservative candidate to win with 40 percent of the vote would not develop.

The moral injunction to be democratic is formal and schematic. How we should be democratic is the substantive question, and an open experimental approach to this question is entirely consistent with an unwavering commitment to democracy. Politicians who are the beneficiaries of a particular and parochial electoral system naturally wrap themselves in the mantle of democracy and need to be reminded occasionally that this mantle can come in different styles, all of them with patches.

ZENO AND MOTION

∞

Around 460 B.C., about eighty years after Pythagoras, the Greek philosopher Zeno of Elea was thinking and writing. Though his writings have not survived, Aristotle's references to them suggest an acute skeptic whose various paradoxes could not be handled by the mathematics of the time. The most famous of these paradoxes concerns the footrace between Achilles and the Tortoise and seems to demonstrate that Achilles, having given the Tortoise a head start, can never catch up to him no matter how fast he runs. In order for Achilles to overtake the Tortoise, he must first reach point T_1, which is the Tortoise's starting point. During this time, however, the Tortoise will have proceeded to point T_2. Now Achilles must hurry over the distance between T_1 and T_2, but while he does so the Tortoise will have inched forward to point T_3. As Achilles covers the distance between T_2 and T_3, the Tortoise will move on a bit to point T_4.

Thus, argued Zeno, Achilles will never pass the Tortoise, because to do so, he will have to complete an infinite number of acts in a finite period of time. That is, Achilles must traverse the distance between T_0 and T_1, between T_1 and T_2, between T_2 and T_3, between T_3 and T_4, between $T_{17,385}$ and $T_{17,386}$, and so on. Since traversing each of these distances takes some time, traversing an infinity of them will take an infinite amount of time. Hence Achilles will never catch up to the Tortoise, although he will get closer and closer to him. The conclusion is clearly false, but what is wrong with the argument?

Before I describe a way out of this dilemma, let me introduce a couple more. In the paradox of the arrow Zeno claimed that an arrow stands still even when it is in midflight. At any particular instant, the arrow simply is where it is and occupies a volume of space exactly equal to itself. During this instant, during this specific moment of time, the arrow cannot move, since, if it did, at least one of two ludicrous consequences would follow. Motion would imply that the instant had an earlier and a later part, and instants by definition don't have parts. Or else it would imply that the arrow would have to occupy a volume of space larger than itself in order for it to have room to change position. Neither of these makes sense, so we conclude that motion during an instant is impossible. We conclude further that motion is impossible, since, were it to occur, it would have to take place during some instant or other.

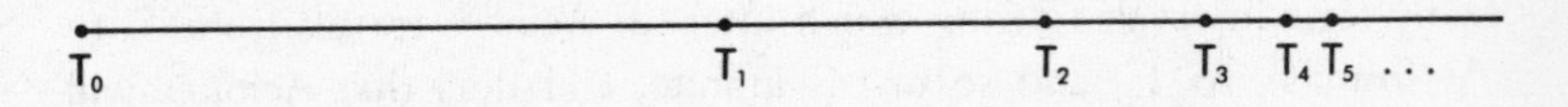

To catch up to the Tortoise, Achilles must traverse the distance from T_0 to T_1, from T_1 to T_2, from T_2 to T_3, and so on. Zeno argued that since traversing each of these intervals takes some time, traversing an infinity of them should take an infinite amount of time.

Again the conclusion is clearly false, but what's wrong with the argument? The genius of Zeno lies in part in his willingness to follow arguments where they lead even if they lead to contradictory positions, in this case regarding the nature of space and time. Are space and time continuous and infinitely divisible? Then how do you account for the Achilles paradox? Are they discrete and jerkily cinematographic? Then how do you account for the arrow paradox?

One last puzzle that doesn't rank with the two previous ones in historical importance but is rather something of a trick: Since it also deals with motion and Greek sporting events, however, it may be appropriate here. Imagine two runners in the senior citizen division competing in a marathon race—-approximately 26.2 miles long. One runner, George, runs at a steady pace of 10 minutes per mile. The other runner, Waldo, runs at a very uneven rate, but it's found that for whatever mile-long stretch along the course he is timed, he always completes the stretch in 10 minutes 2 seconds. The question is: Can Waldo, despite his consistently slower rate, possibly beat George in the marathon?

The answer, of course, is yes, else why pose the problem? What can happen is that Waldo sprints the first .2 mile in 1 minute and then runs the next .8 mile in 9 minutes 2 seconds. He repeats this pattern, alternating 1-minute sprints of .2 mile with more leisurely 9-minute-2-second ambles of .8 mile. He covers every mile including the last one in 10 minutes 2 seconds. Since a marathon is 26.2 miles long, Waldo finishes the race with a sprint. George's total time is 262 minutes (26.2 miles × 10 minutes/mile); Waldo's total time is 261 minutes 52 seconds (26 miles × 10 minutes 2 seconds/mile plus the extra 1 minute for the last .2 mile). Thus Waldo wins by 8 seconds.

And what are the modern resolutions of Zeno's arguments? In the Achilles story, Zeno falsely assumed that the sum total of infinitely many time intervals (during which intervals Achilles travels from T_0 to T_1, from T_1 to T_2, and so on) is infinite, and thus that Achilles will never catch up to the Tortoise. That this isn't necessarily the case can be seen from the series $1 + 1/2 + 1/4 + 1/8 + 1/16 + 1/32 + \ldots$, which, though it contains infinitely many terms, adds up to only 2. It wasn't until the nineteenth century that the rigorization of calculus and infinite series completely clarified these matters (see the entries on *series* and *limits*).

As for the paradox of the arrow, Zeno was right to believe that at any particular instant the arrow is at a particular position. He was also right in believing that there is no intrinsic difference between an arrow being at rest at a particular instant of time and being in motion at that instant; motion and rest are indistinguishable instantaneously. His mistake was in concluding that motion was thus impossible. The difference between rest and motion arises only when we look at positions of the arrow at a number of different time instants. Movement consists in nothing more than being at different places at different times, rest in being at the same place at different times.

A remarkable aspect of the paradoxes is the time lag they demonstrate. Some of the most subtle techniques in calculus and infinite series are needed to clarify puzzles posed 2,500 years ago by this Zeno of Elea. These thought experiments thus provide an extreme example of ideas and narrative preceding equations and calculations.

This is the natural order of development, but too often, especially in mathematical pedagogy, the order is reversed. When this happens, mathematics becomes a collection of techniques, and its intimate relationship with philosophy, literature, history, science, and everyday life is lost. Without this supporting matrix of human meaning and story, mathematics ceases to be one of the liberal arts and becomes merely a technician's tool. I think this tendency should be resisted, and herein is my bit of resistance.

CHRONOLOGICAL LISTING OF THE "TOP FORTY"

Most people like to see a "top forty" list despite the distortions, misapprehensions, and folly such a simplistic notion inevitably produces. At least with record albums there is a more or less clear-cut standard used in constructing the list: the number of records sold within a given interval. No such standard exists for mathematicians scattered over millennia, where the number of theorems proved, the number of citations, the significance and depth of the discoveries, plus a good deal of *je ne sais quoi* all go into the mix. Nevertheless, what follows is a reference list of mathematicians (most have been mentioned in the book) generally considered to be among the "best" from ancient times up to the early twentieth century. I've left out almost all honorific adjectives (greatest, most brilliant) to keep from being repetitive, and I stress the word "among" in the previous sentence.

ANONYMOUS Egyptians, Babylonians, Chinese, Mayans, Indians, and others. The unknown scribes, priests, astronomers, and shepherds who developed the basic notions and notations for number and shape.

PYTHAGORAS (about 540 B.C.). Greek. Along with Thales the founder of Greek mathematics. Pythagorean theorem attributed to his school. Philosopher and number mystic ("All is number").

PLATO (427?–347 B.C.). Greek. Not a mathematician himself, but known as "the maker of mathematicians." His *Dialogues* and *Republic* replete with mathematical speculations and references. Over the doors to his academy inscribed the motto: "Let no one ignorant of geometry enter here."

EUCLID (about 300 B.C.). Greek. Author of *The Elements,* a systematization of Greek geometry, which profoundly affected mathematical thought for millennia.

ARCHIMEDES (287–212 B.C.). Greek. Seminal work in geometry, in particular computing areas and volumes via "method of exhaustions." Greatest scientist of antiquity with discoveries in astronomy, hydrostatics ("*Eureka,* I found it"), and mechanics.

APOLLONIUS (about 230 B.C.). Greek. Further development of geometry in his *Conic Sections.* Anticipated Descartes's analytic geometry. Astronomer.

PTOLEMY (90?–160? A.D.). Alexandrian Greek. Author of the *Almagest* ("the greatest"), a guiding book in astronomy, geography, and mathematics for centuries. Ptolemaic system.

DIOPHANTUS (about 250 A.D.). Alexandrian Greek. Author of *Arithmetica,* dealing with the theory of numbers and the whole-number solution of equations.

AL-KHOWARIZMI (about 830). Baghdadi Arab. His *Al-jabr wa'l Muqabalah* first text in elementary algebra. Very influential in bringing subject to Europe. One of a number of remarkable Arab mathematicians.

OMAR KHAYYÁM (1050?–1123). Persian. Best known as the author of the *Rubáiyát.* Also a notable mathematician whose work in algebra extended that of Al-Khowarizmi. And BHASKARA (1114–1185). Indian. Leading algebraist of twelfth century.

GERONIMO CARDANO (1501–1576) and NICCOLÒ TARTAGLIA (1500–1557). Italian. Discoverers of formulas for solutions to cubic and quartic equations. Stimulated research in algebra.

GALILEO GALILEI (1564–1642). Italian. Although not a mathematician, his *Two New Sciences* helped bring about the revolutionary alliance of mathematics and experimentation and the resulting new methods for pursuing the truths of nature.

FRANÇOIS VIÈTE (1540–1603). French. Invented indispensable notation for algebra, especially variables. Work in algebra and trigonometry. SIMON STEVIN's (1548–1620) decimals, JOHN NAPIER's (1550–1617) logarithms, and JOHANNES KEPLER's (1571–1630) ellipses and astronomical studies date from the same era.

RENÉ DESCARTES (1596–1650). French. Along with Pierre Fermat the inventor of analytic geometry, uniting the two fields of algebra and geometry to create the foundation for modern mathematics. Philosopher, Cartesian doubt.

PIERRE FERMAT (1601–1665). French. Along with Descartes the inventor of analytic geometry. Work in number theory, including his famous "last theorem," in calculus, and in probability.

BLAISE PASCAL (1623–1662). French. One of the founders of probability theory. Pascal's wager. Philosopher, religious mystic, and literary stylist.

ISAAC NEWTON (1642–1727). English. Invented/discovered the binomial theorem, infinite series, and the calculus (simultaneously with Leibniz), from which time modern mathematics may be dated. Also founded modern physics, including laws of motion, gravitation, and optics in his *Mathematical Principles of Natural Philosophy.*

GOTTFRIED WILHELM VON LEIBNIZ (1646–1716). German. Invented the calculus (simultaneously with Newton). Philosopher and logician who anticipated many developments.

The BERNOULLIS, including JAKOB (1654–1705), JOHANN (1667–1748), DANIEL (1700–1782), and others. Swiss family of mathematicians. Important contributions to calculus of variations, probability, and mathematical physics.

LEONHARD EULER (1707–1783). Swiss. Prolific in many areas of mathematics. Significant theorems in calculus, differential equations, number theory, applied mathematics, combinatorics.

JOSEPH LOUIS LAGRANGE (1736–1813). French. Number theory, differential equations, calculus of variations, analysis, celestial mechanics, and dynamics. His *Analytical Mechanics* a beautiful work of pure mathematics.

PIERRE SIMON LAPLACE (1749–1827). French. Influential work in mathematical analysis and astronomy. Established probability as a serious branch of mathematics. Contemporaneous with the MARQUIS DE CONDORCET's (1743–1794) social applications of mathematics and A. M. LEGENDRE's (1752–1833) analysis.

NIKOLAI LOBACHEVSKI (1793–1856) and JÁNOS BOLYAI (1802–1860). Russian and Hungarian, respectively. Discoverers of non-Euclidean geometry (with Gauss), in which Euclid's parallel postulate fails to hold.

KARL FRIEDRICH GAUSS (1777–1855). German. Many deep results in number theory (*Arithmetical Researches*), theory of surfaces, non-Euclidean geometry, mathematical physics, and statistics (bell-shaped curve). Along with Newton and Archimedes, commonly judged to be one of the top three mathematicians of all time.

AUGUSTIN LOUIS CAUCHY (1789–1857). French. Important theorems in complex variables, other work in analysis and group theory. Initiated more rigorous, formal methods in calculus.

ADOLPHE QUÉTELET (1796–1874). Belgian. First to champion the use of probability and statistical models to describe social, economic, and biological phenomena.

EVARISTE GALOIS (1811–1832). French. Work in the theory of equations determined the shape of modern abstract algebra.

WILLIAM HAMILTON (1805–1865), ARTHUR CAYLEY (1821–1895), and J. J. SYLVESTER (1814–1897). Irish, English, and English, respectively. Developers of abstract algebra as the study of operations, form, and pattern. Matrices. Quaternions.

BERNHARD RIEMANN (1826–1866). German. His very original work laid the geometrical framework for Einstein's general relativity theory and liberated geometry from its dependence on the physical notions of length, width, and depth.

GEORGE BOOLE (1815–1864). English. His *Laws of Thought* brought a mathematical approach to the study of logic which was significantly furthered by GOTTLOB FREGE (1848–1925) and GIUSEPPE PEANO (1858–1932).

JAMES CLERK MAXWELL (1831–1879). Scottish. Not a mathematician, but his mathematical development of the theory of the electromagnetic field warrants his inclusion here, as does the mathematical work of the American JOSIAH GIBBS (1839–1903).

GEORG CANTOR (1845–1918). German. Inventor/discoverer of set theory, including the study of infinite sets and transfinite numbers. Also work on irrational numbers and infinite series.

FELIX KLEIN (1849–1925). German. Unified the study of geometry by examining those properties left invariant by a particular group of transformations. Influential teacher.

JULES HENRI POINCARÉ (1854–1912). French. Contributor to many fields of mathematics, including topology, differential equations, mathematical physics, and probability. Also more expository works on scientific method.

DAVID HILBERT (1862–1943). German. Leader of formalist school of axiomatic mathematics and author of *Foundations of Geometry,* which cleaned up logical holes in Euclid. Proposer of list of famous unsolved problems. Versatile contributor to algebra, theory of abstract spaces (Hilbert spaces), and analysis (space-filling curves).

G. H. HARDY (1877–1947), J. E. LITTLEWOOD (1885–1977), and S. RAMANUJAN (1887–1920). English, English, and Indian, respectively. Much original work in analysis and number theory, individually and in collaboration, with Ramanujan relying largely on pure intuition. Some more general writings as well.

BERTRAND RUSSELL (1872–1970). English. Author of *Principia Mathematica* (with Alfred North Whitehead), deriving all of mathematics from logic alone (almost). Russell's paradox. Philosopher and popular author. (Although this is not one of his claims to fame, Russell was a hero of mine in junior high school, and, having included a letter I wrote him while in college in his autobiography, he remains one to a limited extent even today.)

KURT GÖDEL (1906–1978). Austrian/American. His incompleteness theorem established the existence of undecidable propositions within every formalization of mathematics. Other highly original work on consistency, intuitionism, and recursion.

ALBERT EINSTEIN (1879–1955). German/American. Although not specifically mathematical, his revolutionary work on general relativity (not to mention all his other labors) earns him a place on this list.

JOHN VON NEUMANN (1903–1957). Hungarian/American. Seminal contributions to foundations of mathematics, mathematical physics (especially quantum theory), analysis, and abstract algebra. Invented the theory of games. Important work on computers and automata.

This list necessarily omits many great mathematicians of the past. Neither does it contain *any* present-day mathematicians, for several reasons: They are too numerous to mention (there are more than 30,000 names on the combined membership lists of the principal American professional organizations alone and hundreds of research journals); the verdict on their work is not yet clear; their results are sometimes too specialized to even superficially summarize; and there are many peripheral people and fields that might also qualify—Benoit Mandelbrot and his fractals, statisticians such as Ronald Fisher and Karl Pearson, computer scientists Alan Turing, Marvin Minsky, and Donald Knuth, A. N. Kolmogorov and his work in probability and complexity, Claude Shannon on information theory, Kenneth Arrow on social choice functions, and many others.

A quick skimming of the list also reveals that there are no women, blacks, or Orientals on it. Recent talk about expanding the canon of literary works and authors has not extended to the mathematical pantheon. Of course, this is not to say that these groups do not possess mathematical talent of the highest order. From Hypatia, an Alexandrian woman who wrote mathematical commentaries and was killed by a fanatical Christian mob in 415, to Lord Byron's daughter Ada Lovelace, who first wrote programs for Charles Babbage's analytic engine (an elaborate mechanical computer), to Emmy Noether, an eminent algebraist who was dismissed by the Nazis and taught at Bryn Mawr College in the 1930s, women have had to struggle for Virginia Woolf's minimum requirement of a room of one's own. This situation has improved a little, and recent years have seen a significant minority of mathematics Ph.D.'s going to females. (New research indicates that most women mathematicians have had a supportive family—many with fathers in mathematics, exposure to serious mathematics at a young age, and women role models in the field.)

As for Orientals, blacks, and other ethnic groups, one needn't look far for evidence of their mathematical acumen to dispel the historical Eurocentrism of the list. Consider, to name just a few manifestations at random, the long history of Chinese mathematics (not well known in the West—Pascal's triangle having been discovered 300 years before Pascal, for example), the mathematical and quasi-mathematical achievements of many "primitive" cultures, the technical accomplishments of Egyptians, Indians, Africans, and Arabs, and the present-day research of Japanese mathematicians. Whatever the restrictions and parochialism of the past, mathematics is today studied on every continent and in every country in the world. The truism of the universality of mathematics probably won't be fully appreciated, however, until there is a corresponding universality of mathematicians.

SUGGESTED READINGS

For the most part, the following books are accessible to a general reader. They contain few equations and emphasize mathematical ideas, not routine computations or rigorous proofs. A reader with a strong mathematical background will probably already know where else to look for further reading.

Abbot, Edwin A. *Flatland.* Dover Publications, 1952.

Albers, Donald J., and G. L. Alexanderson, editors. *Mathematical People.* Birkhäuser, 1985.

Beckmann, Petr. *A History of Pi.* St. Martin's Press, 1971.

Bell, Eric Temple. *Mathematics: Queen and Servant of Science.* Mathematical Association of America, 1987.

Benacerraf, Paul, and Hilary Putnam, editors. *Philosophy of Mathematics.* Prentice-Hall, 1964.

Boyer, Carl. *The History of Calculus and Its Conceptual Development.* Dover Publications, 1959.

———. *A History of Mathematics.* John Wiley & Sons, 1968.

Chinn, W. G., and N. E. Steenrod. *First Concepts of Topology.* Mathematical Association of America, 1966.

COMAP (Consortium for Mathematics and Its Applications). *For All Practical Purposes: Introduction to Contemporary Mathematics.* W. H. Freeman, 1988.

Courant, Richard, and Herbert Robbins. *What Is Mathematics?* Oxford University Press, 1948.

Daintith, John, and R. D. Nelson, editors. *Dictionary of Mathematics.* Penguin Books, 1989.

Davis, Philip J. *The Lore of Large Numbers.* Mathematical Association of America, 1961.

——— and Reuben Hersh. *The Mathematical Experience.* Houghton Mifflin, 1981.

Delong, Howard. *A Profile of Mathematical Logic.* Addison-Wesley, 1971.

Eves, H., and C. V. Newsom. *An Introduction to the Foundations and Fundamental Concepts of Mathematics.* Holt, Rinehart and Winston, 1965.

Gardner, Martin. *Aha! Insight.* W. H. Freeman, 1978.

———. *Aha! Gotcha.* W. H. Freeman, 1982.

———. *Wheels, Life and Other Mathematical Amusements.* W. H. Freeman, 1983.

———. *Penrose Tiles to Trapdoor Ciphers.* W. H. Freeman, 1989.

Gleick, James. *Chaos: Making a New Science.* Viking Press, 1987.

Guillen, Michael. *Bridges to Infinity.* Jeremy P. Tarcher, 1983.

Hardy, G. H. *A Mathematician's Apology.* Cambridge University Press, 1967.

Heath, T. L. *A History of Greek Mathematics.* Clarendon Press, 1921.

Hofstadter, Douglas. *Gödel, Escher, Bach.* Basic Books, 1980.

———. *Metamagical Themas.* Basic Books, 1985.

Ifrah, Georges. *From One to Zero: A Universal History of Numbers.* Viking Press, 1985.

Kac, Mark, and Stanislaw Ulam. *Mathematics and Logic.* Frederick Praeger, 1968.

Kline, Morris. *Mathematical Thought from Ancient to Modern Times.* Oxford University Press, 1972.

———. *Mathematics: An Introduction to Its Spirit and Use.* W. H. Freeman, 1978.

———. *Mathematics for the Non-mathematician.* Dover Publications, 1985.

Littlewood, J. E. (Bela Bollobas, editor). *Littlewood's Miscellany.* Cambridge University Press, 1986.

Mandelbrot, Benoit. *The Fractal Geometry of Nature.* W. H. Freeman, 1977.

Moore, David S. *Statistics: Concepts and Controversies.* W. H. Freeman, 1979.

Niven, Ivan. *The Mathematics of Choice.* Mathematical Association of America, 1965.

Ore, Oystein. *Graphs and Their Uses.* Mathematical Association of America, 1963.

Packel, Edward. *The Mathematics of Games and Gambling.* Mathematical Association of America, 1981.

Pappas, Theoni. *The Joy of Mathematics.* Wide World Publishing/Tetra, 1989.

Paulos, J. A. *Mathematics and Humor.* University of Chicago Press, 1980.

———. *I Think, Therefore I Laugh.* Columbia University Press, 1985.

———. *Innumeracy: Mathematical Illiteracy and Its Consequences.* Farrar, Straus & Giroux, 1989.

Peterson, Ivars. *The Mathematical Tourist.* W. H. Freeman, 1988.

———. *Islands of Truth.* W. H. Freeman, 1990.

Polya, George. *How to Solve It.* Princeton University Press, 1945.

Poundstone, William. *The Recursive Universe.* William Morrow, 1985.

Rucker, Rudy. *Mind Tools.* Houghton Mifflin, 1987.

Russell, Bertrand. *The Principles of Mathematics.* Cambridge University Press, 1903.

Salmon, Wesley. *Space, Time, and Motion.* Dickenson, 1975.

Steen, Lynn Arthur, editor. *Mathematics Today.* Springer-Verlag, 1978.

Weaver, Warren. *Lady Luck.* Dover Publications, 1982.

Weizenbaum, Joseph. *Computer Power and Human Reason.* W. H. Freeman, 1976.

Wilder, Raymond. *The Foundations of Mathematics.* John Wiley & Sons, 1965.

INDEX

ALSO BY JOHN ALLEN PAULOS

INNUMERACY

Through witty, amusing, illustrative examples, Paulos enlightens readers to the world of mathematics, making numbers, probability, and statistics as easy to understand as ABC.

"An intelligent and good-natured analysis of the follies that arise from a lack of understanding . . . of science and mathematics. This pleasant book would improve the quality of thinking of virtually anyone."

—Isaac Asimov

Nonfiction/0-679-72601-2/$8.95

I THINK, THEREFORE I LAUGH

Modern philosophy meets Groucho Marx in this funny and insightful look into the worlds of truth, logic, reasoning, artificial intelligence, and more. *I Think, Therefore I Laugh* incorporates Wittgenstein's argument that "a serious and good philosophical work could be written that would consist entirely of jokes."

Philosophy/0-679-72954-2/$8.95